DESIGNING SAFER POLYMERS

DESIGNING SAFER POLYMERS

PAUL T. ANASTAS, Ph.D.
PAUL H. BICKART, Ph.D.
MARY M. KIRCHHOFF, Ph.D.
Office of Pollution Prevention and Toxics
U.S. Environmental Protection Agency

A JOHN WILEY & SONS, INC., PUBLICATION

New York • Chichester • Weinheim • Brisbane • Singapore • Toronto

This book is printed on acid-free paper. ⊗

Published simultaneously in Canada.

For ordering and customer service, call 1-800-CALL-WILEY.

ISBN 0-471-39733-4

Printed in the United States of America.

10 9 8 7 6 5 4 3 2 1

CONTENTS

PREFACE

In 1984, the Environmental Protection Agency published a Toxic Substances Control Act (TSCA) section 5(h)(4) rule granting an exemption for the manufacture and importation of certain polymers. The Agency has reviewed thousands of polymers in the interim and has developed internal guidelines for identifying polymers that do not pose an unreasonable risk of injury to human health or the environment. The final rule (USEPA 1995) reflects the Agency's extensive experience in evaluating polymers and expands the 1984 exemption criteria to allow more low-risk polymers to qualify for exemption.

The scope of this book extends beyond the polymer exemption rule itself. The fusion of green polymer chemistry with the regulatory provisions for polymers under TSCA provides a useful reference for industrial scientists and decision makers in the polymer industry. Research scientists in the polymer field will appreciate the diverse topics it addresses and the unique examples it offers.

This book is intended to assist those in the polymer industry in designing safer polymers, substances that adhere to the exemption criteria outlined in the polymer exemption rule. The first chapter recognizes the central role of polymers and plastics in society, while identifying some of the concerns associated with their use and manufacture. Chapter 2 focuses on green chemistry, the design of products and processes that eliminate or reduce the use and generation of hazardous substances. Industry can achieve substantive environmental gains by incorporating the 12 principles of green chemistry into process design and development. The third chapter examines the Toxic Substances Control Act and the exemptions permitted under this legislation, including the polymer exemption. Chapter 4 provides detailed guidance on meeting the polymer exemption criteria. In designing polymers that are exempt, industry is manufacturing products that are better for the environment while simultaneously reducing their regulatory obligations.

P.T.A.
P.H.B.
M.M.K.

Washington, DC

LIST OF TABLES

LIST OF FIGURES

CHAPTER 1

THE ROLE OF POLYMERS IN SOCIETY AND THE ENVIRONMENT

1.1 INTRODUCTION

The growth in the use of polymers and plastics has had remarkable parallels to the growth of the environmental movement. Both of these movements have their origins in the early part of this century; however, it was during the 1960s that a widespread explosion in awareness and adoption of these two new frontiers occurred. Whether it was in 1962 when Rachel Carson penned *Silent Spring* or in 1967 when the film *The Graduate* advised popular culture that the future was just one word, "plastics," the two areas have been growth industries ever since, and, since their origins, they have tracked one another in other aspects as well.

Many of the earliest concerns for environmental problems focused on those that were most visible, such as litter and solid waste. As consciousness about the issue rose, so did a desire to change people's littering behavior along with the very materials that constitute the litter. It was during this period that the polymer and plastics industry first began to focus on designing materials that biodegrade in the natural environment. Efforts continue to this day to make a wide range of synthetic materials that do not persist in the environment intact but break down into innocuous degradation products.

As environmental awareness grew, so did concerns for more complex problems such as water quality. As a result, the field of polymer science again offered some innovative technologies that helped make the water quality of today far superior to that of a generation ago. Whether through the use of water purification systems involving flocculants or through the development of complex filters and membranes, the polymer and plastics industry has responded to water-based environmental issues in a visionary way.

Today, with greater understanding of the root cause of threats to human health and the environment, the environmental movement and the underlying science are focusing on avoiding problems rather than cleaning them up after the fact. This new focus on pollution *prevention* and a method to achieve it is known as green chemistry (see Chapter 2). This approach is applicable to the field of polymers and plastics, and the scientific and industrial communities involved with polymers are responding with vision and innovation.

This book aims to provide a twofold perspective. The first is to illustrate the general approach to designing polymers that reduce and eliminate the use and generation of hazardous substances throughout all stages of the polymer life cycle. The second is to relate this approach to the current primary environmental regulations in place under the Toxic Substances Control Act (see Chapter 3), the Polymer Exemption. By using and understanding the criteria outlined in the regulatory framework, one can generate general guidelines that provide a blueprint for designing substances that not only reduce the impact on human health and the environment but also reduce the regulatory burden on manufacturers.

1.2   FROM THE STONE AGE TO THE POLYMER AGE

It is difficult to imagine our society without polymers. From plastic soda bottles to automobile parts, from medical implants to bulletproof vests, polymers are an integral part of our daily lives. New polymers and new uses for existing polymers are constantly being developed. Fully half of the new chemicals entering commerce in the United States each year are polymers. In addition ethylene, primarily used as a monomer for plastics, is second only to sulfuric acid in production volume.

The plastics industry plays a major role in the economy of the United States and the world. In 1996, shipments of plastics in the United States totaled $274.5 billion, a 55% increase since 1991 (1). The industry employed more than 1.3 million workers in 1996. Globally that same year, 95.6 million metric tons of the five leading thermoplastic resins were consumed: high-density polyethylene (HDPE), low-density polyethylene (LDPE/LLDPE), polypropylene, polystyrene, and polyvinyl chloride (PVC).

The versatility of polymers has been central to their extensive use in a wide variety of applications. For example, consider the use of plastics as beverage containers. Eight gallons of juice can be delivered by 2 pounds of plastic, 3 pounds of aluminum, or 27 pounds of glass (2). The strength of plastics allows less packaging to be used, decreasing production, shipping, and energy costs. Improvements in plastics themselves have resulted in soda bottles and milk jugs that use less material today than used 20 years ago.

In addition to consumer products, plastics perform a critical function in life-saving devices used in hospitals and emergency rooms. Syringes are used to deliver essential drugs, IV bags dispense vital fluids, and plastic tubing transfers oxygen to assist with breathing. The inertness of plastics makes them ideal candidates for use in prosthetic limbs and pacemakers. Biodegradable stitches are commonly used to eliminate the need for subsequent removal. Improved medical devices and procedures have benefited from better polymer products.

1.3 CHARACTERISTICS OF POLYMERS AND THE ASSOCIATED CONCERNS

Polymers share several common features that distinguish them from other molecules, such as high molecular weights and the presence of repeating units. A high molecular weight limits transport across biological membranes; therefore, a polymer with a high molecular weight may not pose a threat to human health or the environment through absorption. A typical polymer sample, however, contains polymer molecules that exhibit a range of molecular weights. This lack of uniformity necessitates reporting the molecular weight of the sample as either the number-average molecular weight (M_n) or the weight-average molecular weight (M_w). The lower molecular weight species present in a polymer sample may be transportable across biological membranes, presenting a potential concern to living organisms.

The polymerization process itself decreases the number of reactive functional groups present in the final polymer. The bond-forming reactions incorporate the functional groups of the monomers into the polymer backbone, often leaving reactive functionality only as end groups. These polymers are more inert than the starting monomers, decreasing potential health and environmental problems. However, some polymers are synthesized with reactive functional groups imbedded in or pendant to their backbone. The presence of such groups facilitates cross-linking between polymer chains, reactions with other molecules, or complexation with added reagents. Increased toxicity may be a concern when certain reactive functional groups are present.

A potential hazard to human health and the environment is posed by macroscopic properties of polymers as well as microscopic properties. Consider, for example, a polymer that is inhaled and leads to respiratory distress because it fails to clear the lungs. The biological effects may have less to do with the molecular structure than with the gross physical features. Problems associated with macroscopic properties are evident in the well-publicized case of the environmental persistence of plastic six-pack rings. The failure of this product to degrade created a hazard to marine life that became entangled in the ring. Consumers were encouraged to cut up six-pack rings before disposal to eliminate the threat to wildlife. An alternative approach to protecting marine life would entail designing a six-pack ring that degrades in the environment.

A key positive characteristic of polymers in a variety of applications, however, is their durability. This feature, which makes polymers so versatile, also creates a problem due to persistence in the environment. Many polymer products do not degrade in landfills or sewage treatment plants. Recycling of plastics helps to minimize the disposal problem; however, not all plastics are currently capable of being recycled. The design of biodegradable polymers demonstrates industry's practical approach to addressing the issue of persistence in the environment.

Another important feature is the use of various additives. In many cases, product performance may be optimized by combining additives with the polymer during the processing stage. Fire retardants, for example, may be added to clothing. Antioxidants may be used to increase the durability of a product, and dyes can be added to give the polymer color. Although additives enhance polymer properties, they can also leach from the polymer product into the surrounding environment. The polymer and plastics industry continues to identify new additives that give the same desirable properties without posing a threat to human health and the environment.

1.4   CHARACTERISTICS OF POLYMER MANUFACTURE AND SYNTHESIS

Environmental concerns extend beyond the polymer products themselves, to synthesis and manufacturing. These processes traditionally have utilized nonrenewable feedstocks, volatile organic solvents, or hazardous reagents. In some cases, the reaction may require a large input of energy or may generate a significant quantity of waste.

Most polymer syntheses, like 98% of all organic chemical products, begin with petroleum-based feedstocks. Concerns for the dwindling supply of oil necessitates the development of alternative feedstocks. Many of the starting materials currently in use are also of concern because of carcinogenicity. Consequently, utilization of biomass presents an alternative to current feedstocks that is both renewable and relatively innocuous to human health and the environment.

Polymer synthesis often requires the use of large amounts of organic solvents. Numerous problems are associated with the use of these solvents: toxicity, volatile organic compound (VOC) emissions, disposal, and contamination of the aqueous waste stream. These solvents are often highly regulated under the Clean Air Act, the Clean Water Act, and the Toxics Release Inventory. Compliance with these regulations can be very costly to a company. Polymerization reactions that use an aqueous environment or no solvent at all may be designed, as in a solid-state polymerization. Environmentally friendly solvents, such as supercritical CO_2, may also be appropriate.

Polymer properties are fine-tuned through the addition of auxiliary substances in a process known as compounding. These additives are used to improve a polymer's stability, processability, or physical properties. Compatibilizers, for example, allow two or more materials, such as virgin and post-consumer plastic resins, to coexist indefinitely (3). Heat stabilizers reduce thermal degradation during processing and product lifetime. Fillers may be incorporated into the polymer to improve strength and thermal stability. Flame retardants are essential when processed into polymers that are used in confined spaces, such as airplanes. Antioxidants prevent breakdown of the polymer through oxidation; phenol derivatives are commonly used in this capacity. Plasticizers improve the flexibility of the polymer by lowering the glass transition temperature below room temperature. Unfortunately, these additives can leach out of the polymer product and into the surrounding environment. Recently, concern has been raised about the potential leaching of phthalates, used as plasticizers in children's toys. The additive that makes a teething ring soft and pliable may present a health risk to the infant chewing on it.

Plastics additives were estimated to be a $16 billion industry in 1997 (4). Modifiers, which improve impact strength or flexibility, are the largest category of additives. Plasticizers, impact modifiers, and organic peroxides provide examples of modifiers. Property extenders, such as antioxidants, light stabilizers, and flame retardants, impart stability during processing and use. Processing aids improve the moldability of a plastic and include mold-release agents and lubricants.

A polymer that, by itself, may be considered safe, may contain comonomers and additives with potentially harmful effects. For example, polyvinyl alcohol is approved by the Food and Drug Administration (FDA) for use in cosmetics and food wraps yet is manufactured with substances that present a health risk to plant employees (5). Comonomers used in the synthesis of polyvinyl alcohol include acrylamide, a neurotoxin; formaldehyde, a suspected carcinogen and powerful mutagen and teratogen; and vinyl chloride, a known human carcinogen. The cross-linking agent, divinyl sulfone, is highly toxic, causing skin and eye burns similar to those caused by mustard gas. Materials used in manufacturing and processing also need to be considered when designing safer polymers.

Numerous polymerization reactions utilize hazardous reagents, such as phosgene. The transportation and handling of such reagents may require special equipment, clothing, and safety precautions. These additional safeguards increase the costs associated with the manufacturing process and have the potential to fail. New and existing technologies should be examined to identify safer reagents to those currently in use.

The polymer manufacturing process often requires a large energy input. A prolonged heating period may be necessary to effect polymerization; cooling may be required to maximize product yield; energy-intensive separations may be needed to isolate the desired product. High energy requirements further deplete existing fossil fuels. Ideally, polymer processing and manufacture should be conducted at ambient temperature and pressure. Catalysts have proven especially useful in minimizing the energy needed to carry out specific transformations.

It is not uncommon for batches of polymers to be "off-spec," resulting in a product that may be commercially unusable. Unless another use is found for this off-spec material, this substandard product will become part of the waste stream. This problem can be minimized by increasing the use of in-line monitoring. Continuous feedback on the reaction progress and polymerization conditions permits real-time process adjustments, decreasing the probability of producing substandard polymers.

A final factor to consider in polymer manufacturing is atom economy, the efficiency with which atoms in the starting materials are incorporated into the final product. Polymerization reactions frequently demonstrate high atom economy. Addition polymerization adds monomers and reactants without loss of mass, whereas condensation polymerization yields products with loss of a small molecule, such as water. Addition reactions inherently demonstrate higher atom economy than elimination reactions, thereby decreasing the number of atoms contributing to the waste stream.

1.5 CONCLUSION

The polymer industry has revolutionized the quality of life through the generation of new materials that have transformed everything from transportation to communication to recreation. However, it is also recognized that there are still challenges to meet in the development of new polymers and polymer-manufacturing technologies. Through the discovery, development, and implementation of innovations to reduce hazardous substances in all stages of the polymer life cycle, the polymer industry is achieving its dual goals of both protecting human health and the environment and being more profitable by reducing a variety of costs, for example, regulatory costs. These techniques are finding widespread use and will continue the traditional parallels between the fields of environmental protection and polymer science.

REFERENCES

1. http://ameriplas.org/benefits/economic/economic.html
2. http://ameriplas.org/benefits/about_plastics/uses.html
3. http://ameriplas.org/top_level/glossary/glossary_c.html
4. *Chem. Eng. News* 1998. 76 (40): 22.
5. Radian Corporation. 1986. *Polymer Manufacturing: Technology and Health Effects,* p. 550. Park Ridge, NJ: Noyes Data.

CHAPTER 2

GREEN CHEMISTRY AND ITS ROLE IN DESIGNING SAFER POLYMERS

Green chemistry is the design of chemical products and processes that reduce or eliminate the use or generation of hazardous substances (1). Through careful planning, chemists are able to manipulate molecules and develop products and processes that minimize harmful effects to human health and the environment. Green chemistry employs the fundamental principles of chemistry in reducing pollution at its source. It embraces chemistry as the solution to many environmental problems.

Green chemistry represents a paradigm shift from pollution remediation, exemplified by the "command and control" approach, to pollution prevention. Historically, environmental problems have been addressed after they have occurred; waste treatment, for example, occurs at the end of the pipe. By preventing pollution, green chemistry eliminates the need for end-of-the-pipe treatment. The Pollution Prevention Act of 1990 (2) validated prevention as the premier approach to environmental protection. Environmental Protection Agency (EPA) Administrator Carol Browner has reiterated pollution prevention as the EPA's "central ethic" (3).

2.1 BACKGROUND

Chemistry is often perceived as being both beneficial and detrimental to human health and the environment. Chemistry has provided products that have improved quality and duration of life, from polymers to textiles to pharmaceuticals. Antibiotics destroy bacteria that once were fatal; vaccines have eradicated diseases that previously caused epidemics; air conditioning prevents deaths in the elderly during prolonged heat waves.

Accompanying these advances, however, have been unwanted and unanticipated side effects—chemical accidents, contamination of air and water, destruction of ecosystems, and depletion of the ozone layer by chlorofluorocarbons (CFCs). Designing safer chemicals and chemical reactions can minimize the unintended consequences of technological advances.

The effort to minimize the impact of pollution on human health and the environment has resulted in an explosion in the number of environmental regulations in recent years (Fig. 2.1). Public awareness of environmental problems has been heightened by well-known incidents, such as the birth of thalidomide babies in Europe in the 1960s; the dumping of chemicals in Love Canal, New York, and Times Beach, Missouri; and the decline in the bird population due to widespread use of DDT.

The deadly methyl isocyanate gas leak in Bhopal, India, in 1984 promoted the passage of the Emergency Planning and Community Right-to-Know Act (EPCRA) by the United States Congress in 1986 (4). EPCRA provides citizens and communities with information related to chemical hazards in their area. Data on hazardous substances are compiled in the EPA's Toxics Release Inventory (TRI). This database tracks the fate of 644 chemicals, a small fraction of the more than 75,000 chemicals in commercial use in the United States. The majority of chemicals on the TRI list are released to the air, as shown in Figure 2.2. An even larger quantity of chemical waste is processed on- or off-site for recycling, treatment, disposal, and energy recovery.

The increase in environmental laws has been accompanied by the added cost of compliance with these regulations. For some companies, the cost of compliance equals its research and development budget, which is not the best use of a company's resources. Pollution prevention through green chemistry is an economically sound strategy for allocating a company's resources.

Environmental regulations aim to prevent disasters and ensure a safer workplace. However, regulations can simply minimize risk, but they cannot eliminate it. Risk is a function of both hazard and ex-

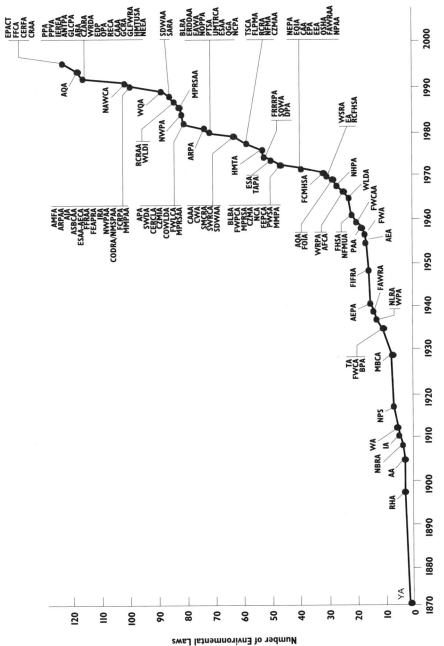

Figure 2.1 Growth in environmental regulation.

11

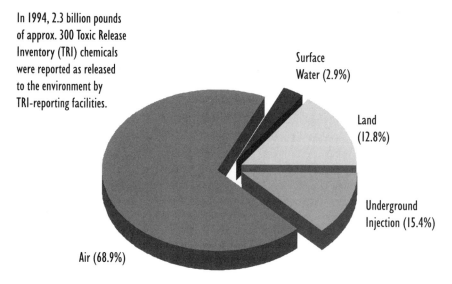

In 1994, 2.3 billion pounds of approx. 300 Toxic Release Inventory (TRI) chemicals were reported as released to the environment by TRI-reporting facilities.

Surface Water (2.9%)

Land (12.8%)

Underground Injection (15.4%)

Air (68.9%)

Figure 2.2   Fate of TRI chemicals released to the environment.

posure. The traditional approach to risk management has been to limit exposure: given the hazard inherent in a particular substance, risk can be minimized by restricting concentration or exposure time. Green chemistry advocates a different approach to risk management. Exposure becomes immaterial if the substance poses no hazard. The use and generation of innocuous substances remove hazard from the equation.

2.2   PRINCIPLES OF GREEN CHEMISTRY

Twelve principles of green chemistry have been identified by Anastas and Warner (1) (Fig. 2.3). These principles serve as a blueprint for designing chemical products and processes that are inherently safer. In many instances, a green chemistry initiative may address several of the principles, resulting in added environmental benefits. It is important to note that trade-offs may be necessary in achieving the goals of green chemistry. For example, a new catalyst may yield a product with higher atom economy and percent yield than a traditional catalyst

1. It is better to prevent waste than to treat or clean up waste after it is formed.
2. Synthetic methods should be designed to maximize the incorporation of all materials used in the process into the final product.
3. Wherever practicable, synthetic methodologies should be designed to use and generate substances that possess little or no toxicity to human health and the environment.
4. Chemical products should be designed to preserve efficacy of function while reducing toxicity.
5. The use of auxiliary substances (e.g., solvents, separation agents, etc.) should be made unnecessary wherever possible and innocuous when used.
6. Energy requirements should be recognized for their environmental and economic impacts and should be minimized. Synthetic methods should be conducted at ambient temperature and pressure.
7. A raw material of feedstock should be renewable rather than depleting wherever technically and economically practicable.
8. Unnecessary derivatization (blocking group, protection/deprotection, temporary modification of physical/chemical processes) should be avoided whenever possible.
9. Catalytic reagents (as selective a possible) are superior to stoichiometric reagents.
10. Chemical products should be designed so that at the end of their function they do not persist in the environment and break down into innocuous degradation products.
11. Analytical methodologies need to be further developed to allow for real-time, in-process monitoring and control prior to the formation of hazardous substances.
12. Substances and the form of a substance used in a chemical process should be chosen so as to minimize the potential for chemical accidents, including releases, explosions, and fires.

Figure 2.3 Twelve principles of green chemistry.

but may be a more toxic catalyst. The synthetic chemist must balance the advantages and disadvantages when designing a reaction.

The twelve principles of green chemistry are briefly discussed below and include examples that illustrate the application of each principle. Although the examples are not perfect, they represent an improvement over existing processes.

Principle 1: It is better to prevent waste than to treat or clean up waste after it is formed

Waste prevention makes sense from both an environmental and economic standpoint. Treatment and disposal of chemical waste pose potential hazards to human health and the environment from accidental spills, leaks, and releases. The cost of separating, treating, and disposing of waste products is enormous. Eliminating waste, such as unreacted starting material, unwanted by-products, and contaminated solvents, from a chemical reaction removes the treatment and disposal steps at the end of the process.

Eastman Kodak has decreased the amount of waste generated in the synthesis of polycarbonate/polydimethylsiloxane copolymers by 600,000 lb/year (5). The new process (Fig. 2.4) avoids the use of bisphenol A bischloroformate, a reactant that is synthesized from phosgene. The aqueous waste stream produced is less hazardous, as triethylamine replaces pyridine as the acid acceptor. The copolymer produced retains the same properties as the original.

The Nalco Chemical Company has developed a new water-based process for manufacturing liquid polymers (6). High molecular weight, water-soluble polymers are commonly used to remove contaminants from industrial waste water. Traditionally, these polymers have been produced as a dry powder or as a water-in-oil emulsion. The powder form presents exposure hazards and is energy intensive to produce and use, whereas the emulsion employs large quantities of hydrocarbon solvents and surfactants. Nalco's new process manufactures water-based acrylamide polymers in an aqueous environment, eliminating the use of oils and surfactants. In addition, ammonium sulfate, a waste by-product in the synthesis of caprolactam, is used in the manufacturing process. Less waste is generated by avoiding the use of oils and surfactants in the manufacture of water-soluble polymers and by recycling a waste product (ammonium sulfate) from another industrial process.

Principle 2: Synthetic methods should be designed to maximize the incorporation of all materials used in the process into the final product

Traditionally, the measure of the success of a chemical reaction has been the percent yield of the desired product. Although yield is im-

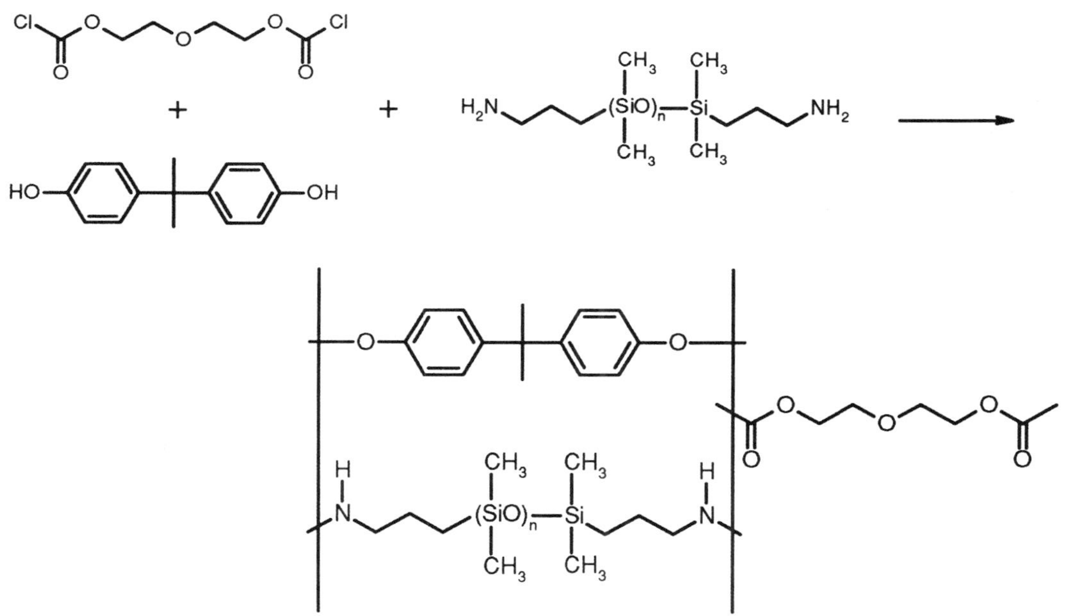

Figure 2.4 Synthesis of polycarbonate/polydimethylsiloxane copolymers.

portant, it does not take into account the fate of all the atoms partic-ipating in the reaction. The concept of atom economy or atom utili-zation, as proposed by Trost (7) and Sheldon (8), emphasizes the importance of incorporating the maximum number of atoms into the final product as a way of minimizing waste. Atoms that are not con-tained within the final product become part of the waste stream. For example, the Diels–Alder reaction is 100% atom efficient; all of the atoms in the starting materials are assimilated into the final product. Elimination reactions, however, are inherently wasteful, as an un-wanted by-product is always formed.

Carbon dioxide has been used as both comonomer and solvent in the synthesis of polycarbonates (Fig. 2.5). Coates and co-workers (9) have designed a faster, more efficient zinc-based catalyst in the poly-merization reaction between cyclohexene oxide and CO_2. The new catalyst features bulky hydrocarbon arms tethered to zinc through nitrogen atoms. Previous catalysts also employed bulky hydrocarbon arms, but these were linked to zinc through oxygen atoms. These Zn–O linkages were susceptible to breaking, causing the zincs to clump together and terminate the reaction. Replacing oxygen with nitrogen provides a stronger bond to zinc, preventing degradation of the catalyst. The improved catalyst is effective under milder reaction conditions and works over 50 times faster than previous catalysts. The metal-catalyzed reaction exhibits 100% atom economy while em-ploying a benign solvent.

Ni(0) catalyzes the polymerization of carbon dioxide with diynes to produce poly(2-pyrones) (10). In this case, CO_2 functions as a

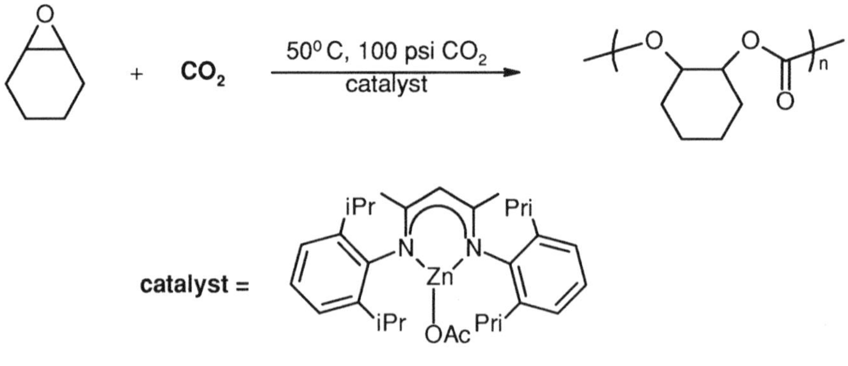

Figure 2.5 Polymerization of cyclohexene oxide with carbon dioxide.

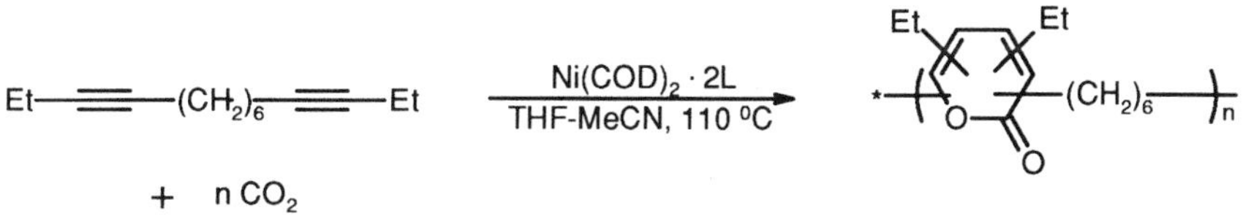

Figure 2.6   Synthesis of poly(2-pyrones).

copolymer but not as solvent (Fig. 2.6); both THF and THF-CH$_3$CN have been utilized as solvents.

Principle 3: Wherever practicable, synthetic methodologies should be designed to use and generate substances that possess little or no toxicity to human health and the environment

Because risk is a function of both hazard and exposure, it can be managed by minimizing the hazardous nature of a substance or by limiting the exposure time or concentration of a substance. Limiting exposure may require the use of special equipment or protective clothing, safeguards that increase the cost of a given process. Exposure controls also have the potential for failure, thereby increasing risk to human health and the environment. Risk management is better addressed by focusing on the hazard of a particular substance. A benign reagent or product presents no risk to human health or the environment regardless of the length of exposure. Obviously, it is desirable to use nontoxic substances wherever possible.

Polyurethane foam has typically been manufactured using CFCs and HCFCs as blowing agents, substances that contribute to ozone depletion and global warming. The Stepan Company has designed a new process that utilizes water as the blowing agent in the manufacture of rigid polyurethane foam (11). During the blowing process, isocyanate and water combine to form polymer and carbon dioxide gas, which becomes trapped in the cellular structure of the foam product. The improved process eliminates the use of ozone-depleting substances while yielding a product with properties that are virtually identical to those of the polyurethane foam manufactured following the traditional procedure.

Sequa Chemicals uses starch graft polymers as phenolic resin extenders to decrease formaldehyde emissions (12). Aqueous phenolformaldehyde (PF) resins contain approximately 2% free formaldehyde, which is released into the atmosphere and workplace each year. Replacing 50% of the phenol-formaldehyde resin with starch grafts cuts formaldehyde emissions in half. The extended resin exhibits improved properties over the original PF resin. The starch graft polymers are water-based, nontoxic, and nonirritating and utilize a renewable resource, starch, in their synthesis.

Principle 4: Chemical products should be designed to preserve efficacy of function while reducing toxicity

A better understanding of the mechanism of toxicity allows structural changes to be made that do not affect the function of a product. For example, toxicity may be avoided by eliminating or modifying a functional group. Physical properties may be altered to minimize bioavailability. A parallel may be found in the pharmaceutical industry, where the effectiveness of a drug is balanced against side effects. In some cases, simply changing a functional group on an aromatic ring may decrease an undesirable side effect yet not diminish the efficacy of the pharmaceutical agent. Structural modifications that decrease toxicity while maintaining function are highly desirable. The chemical industry needs to examine products currently in use for less toxic alternatives.

Coating metals with paint helps to prevent corrosion, a significant problem that costs industry and consumers millions of dollars each year. To remain effective, coatings must be continually stripped and reapplied. Organic solvents, notably methylene chloride, are commonly employed in the stripping process. Because methylene chloride is a known carcinogen, dry-blasting has gained acceptance as a suitable alternative. Raghavan has identified laevo-polylactic acid (l-PLA) as an appropriate dry-blast medium (13). The polymer can be chemically depolymerized after depainting. PLA can be converted to ethyl lactate, a solvent with potential applications in the electronics industry. This technology reduces the amount of solid waste produced by the dry-blast process by 90–97%.

Asahi Chemical Industry Company has developed a solid-state process for the synthesis of polycarbonates (Fig. 2.7) (14). Traditional

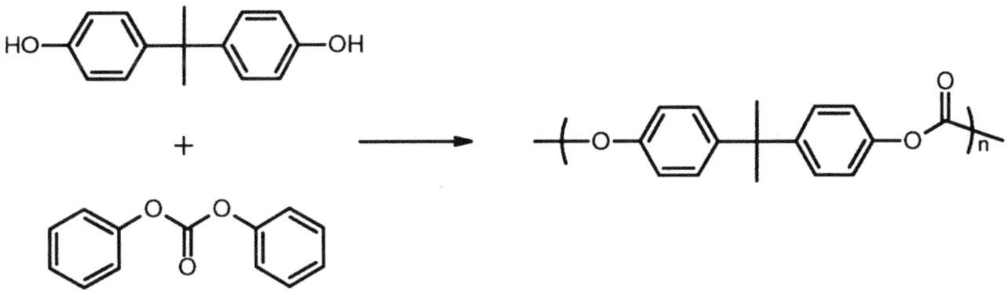

Figure 2.7   Nonphosgene synthesis of polycarbonates.

polycarbonate synthesis employs both phosgene and large quantities of methylene chloride as solvent. Asahi's new process combines bisphenol A with diphenyl carbonate to yield a prepolymer. Diphenyl carbonate itself is also synthesized without the use of phosgene. The crystallized prepolymer is heated under a stream of inert gas in the polymerization step. Higher quality polycarbonates are realized as a result of the solid-state process.

Principle 5: The use of auxiliary substances (e.g., solvents, separation agents, etc.) should be made unnecessary wherever possible and innocuous when used

A major source of waste in the chemical industry is the use of solvents in running a reaction and separating products. Many organic solvents, such as benzene and methylene chloride, are known or suspected carcinogens. Volatile organic compounds (VOCs) are associated with the generation of smog; chlorofluorocarbons (CFCs) have been cited for their role in stratospheric ozone depletion. Although organic solvents may be needed to facilitate a reaction, less hazardous solvents are now being utilized and are proving effective in mediating organic reactions. Benign alternatives to organic solvents include supercritical CO_2 and water; some reactions are even run in the absence of a solvent. The need for separation agents can be minimized by increasing selectivity, often through the use of catalysis.

Buelow and co-workers (15) have polymerized 1,3-butadiene in supercritical CO_2 using a nickel catalyst (Fig. 2.8). The *cis*-1,4 product was highly favored, exhibiting a slight dependence on pressure. Supercritical CO_2 replaces toluene or heptane as the solvent in the polymerization reaction.

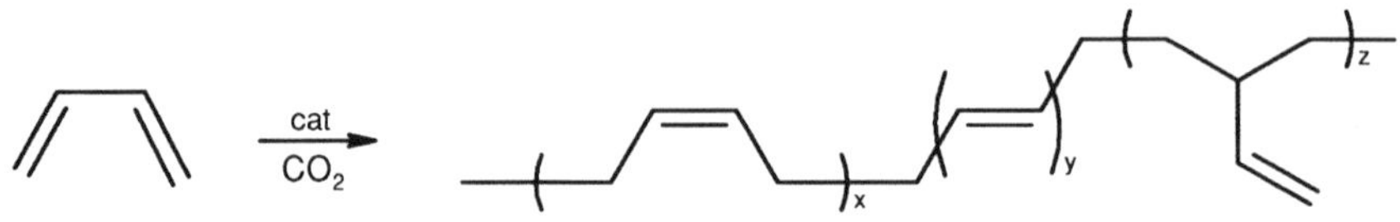

cat = [(π-allyl)Ni(CF$_3$CO$_2$)]

Figure 2.8   Polymerization of 1,3-butadiene in supercritical carbon dioxide.

Mistele and DeSimone (16) have conducted numerous polymerization reactions in supercritical CO_2. For example, supercritical CO_2 is an effective solvent in the synthesis of fluoropolymers, materials that are traditionally synthesized in ozone-depleting CFCs. 1,1-Dihydroperfluorooctyl acrylate (FOA) has been copolymerized with a variety of hydrocarbon-based monomers, such as styrene and methyl methacrylate, in supercritical CO_2 (Fig. 2.9) (17). The homogeneous reaction proceeds via a free-radical process using azobisisobutyronitrile (AIBN) as initiator. When supercritical fluids are used as solvents, the same high-quality products may be synthesized without the use of environmentally damaging solvents.

Principle 6: Energy requirements should be recognized for their environmental and economic impacts and should be minimized. Synthetic methods should be conducted at ambient temperature and pressure

The energy demands of a reaction represent hidden economic and environmental costs. Heating, cooling, separation, and purification of

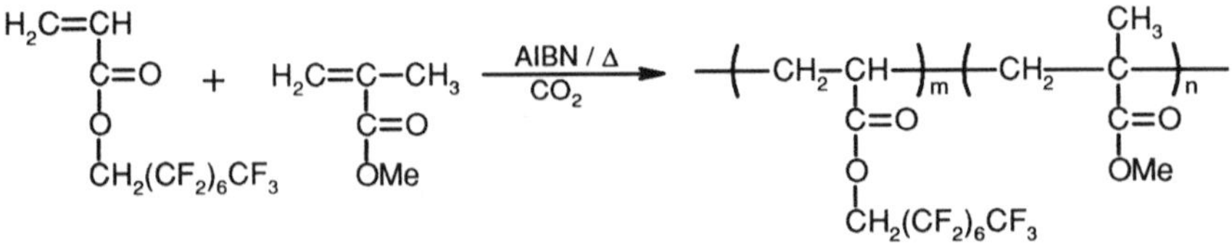

Figure 2.9   Polymerization of 1,1-dihydroperfluorooctyl acrylate with methyl methacrylate in supercritical carbon dioxide.

products require energy in one form or another. Typically, the energy used is supplied by fossil fuel combustion, a process that contributes to numerous environmental problems while depleting a limited natural resource. One approach to minimizing energy requirements is through catalysis, which lowers the activation energy of the reaction, thereby decreasing the amount of thermal energy required to complete the transformation.

The General Electric Corporation has developed an improved catalytic process for the synthesis of its ULTEM polyetherimide resin (18). The new process requires 25% less energy to produce each pound of resin when compared with the old procedure. Additional environmental benefits have also been realized: the volume of the organic waste stream for off-site disposal has been decreased by 90%, the volume of the water-based organic waste for on-site thermal oxidation has been reduced by 60%, the amount of catalyst used has been decreased by 50%, and the amount of waste generated in the synthesis of the catalyst itself has been reduced by 75%. The new synthetic protocol yields both environmental and economic advantages.

Thermal polyaspartic acid (TPA) is a biodegradable polymer that is a suitable substitute for polyacrylic acid (PAC), a nonbiodegradable polymer. The Donlar Corporation has designed a catalytic synthesis of this polymer (19). Use of a catalyst allows the polymerization to occur at a lower temperature and produces a product with superior performance characteristics. Donlar has developed a second synthesis of TPA that employs no organic solvents and produces water as the only by-product.

Principle 7: A raw material of feedstock should be renewable rather than depleting wherever technically and economically practicable

The finite supply of fossil fuels, the primary feedstock for the chemical industry, has spurred interest in developing alternative, renewable feedstocks. Attention has centered on plant-based feedstocks as viable options to petroleum products. Currently, it is difficult for biomass to compete economically with petroleum. As oil becomes more scarce, however, the price of petroleum-based products is expected to rise, making alternative feedstocks economically competitive.

Natural polyhydroxyalkanoates (PHAs) are microbial polyesters that are synthesized by bacteria. Two approaches to PHA biosynthesis have been employed: production by recombinant plant systems and fermentation in aqueous media. Renewable resources, such as fructose and glucose, can serve as the carbon source in the fermentation process. Gross and co-workers have found poly(ethylene glycol) to be effective in controlling both molecular weight and composition of PHAs (20). These polymers have the added advantage of being biodegradable.

Some products may be converted back to feedstock at the end of their useful lifetime. DuPont has designed the Petretec process to recycle polyethylene terephthalate (PET) (21). As shown in Figure 2.10, PET can be broken down into dimethyl terephthalate and ethylene glycol, which can then be used to synthesize PET intermediates. This process decreases the amount of petrochemical feedstocks needed to synthesize PET and decreases the amount of scrap PET that is land filled.

Principle 8: Unnecessary derivatization (blocking group, protection/deprotection, temporary modification of physical/chemical processes) should be avoided whenever possible

It is common practice, particularly in multistep syntheses, to mask sensitive functional groups with protecting groups. Although this prevents unwanted side reactions from occurring, it also adds additional steps that carry with them added use of reagents, solvents, and energy. Converting a product to its salt for ease of separation is a process

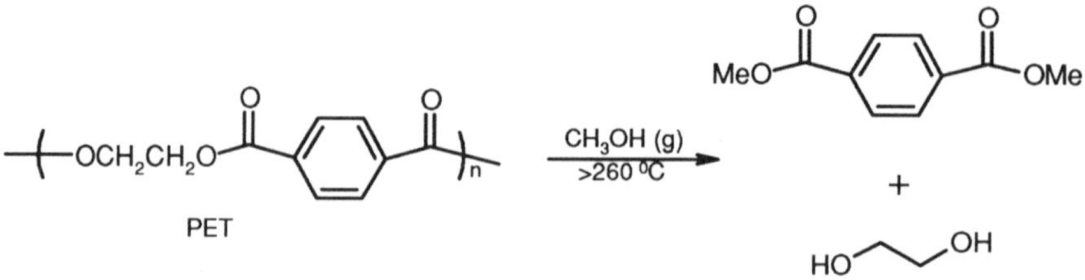

Figure 2.10   Depolymerization of polyethylene terephthalate (PET).

that also requires additional materials. Minimizing the number of unnecessary steps will decrease the amount of waste that is generated.

β-Substituted β-lactones serve as monomers in the synthesis of poly(β-hydroxyalkanoates), PHAs, a class of natural polyesters cited in the previous section. Synthesis of enantiopure monomers requires multistep syntheses that employ protection/deprotection steps. Gross and co-workers (22) have reported the lipase-catalyzed resolution of β-lactones, providing enantioenriched monomers for the ring-opening polymerization reaction. The chemoenzymatic resolution of the β-lactones offers a more efficient route to the starting monomers.

Principle 9: Catalytic reagents (as selective as possible) are superior to stoichiometric reagents

Catalysis offers distinct advantages over stoichiometric reactions. A catalyst may provide enhanced selectivity, eliminating the necessity of separating reaction products. Catalysis may allow a reaction to be performed in a benign solvent, rather than an organic one. Less energy may be required to carry out a catalyzed reaction.

Many advances in polymer chemistry have been achieved through the development of new catalysts. Metallocene, or single-site, catalysts permit the synthesis of polyolefins with improved physical properties (23). These catalysts typically feature a metal in a restricted environment, allowing single access by monomers to this catalytically active site. As a result, polymers grow by a single mechanism, forming a more uniform and reproducible polymer structure. Metallocene-catalyzed polyolefin synthesis yields a product with increased strength and toughness, better clarity and gloss, and improved handling characteristics. Linear low-density polyethylene (LLDPE), high-density polyethylene (HDPE), and polypropylene have all been manufactured with metallocene catalysts.

Swift (24) has designed poly(aspartic acid) to degrade in a sewage treatment plant. The polymer has been synthesized both with and without a catalyst. The acid-catalyzed process yields a more linear polymer that exhibits 90–100% biodegradation in sewage treatment plants. Without a catalyst, the polymer is more highly branched and undergoes less degradation. Poly(aspartic acid) is a replacement for poly(acrylic acid), which is neither biodegradable nor water soluble.

Principle 10: Chemical products should be designed so that at the end of their function they do not persist in the environment and break down into innocuous degradation products

An ideal chemical product would break down into benign substances at the end of its useful lifetime. The persistence of plastics in the environment is well known, and much of the research on biodegradable materials has centered on the polymer industry. Identification of the decomposition products is essential, as the possibility exists that the products could be more harmful than the original substance.

Eastman Chemical Company has developed a copolyester that degrades to water, carbon dioxide, and biomass (25). The polymer consists of adipic acid, terephthalic acid, and 1,4-butanediol, exhibits tensile properties similar to low-density polyethylene (LDPE), and may be blended with natural polymers, such as starch. The compost that results from biodegradation is a soil amendment that adds water retention to soils. Biodegradable polymers such as this one also decrease the amount of waste sent to landfills and incinerators.

Another example of a biodegradable polymer is provided by Mater-Bi, a resin developed by Biocorp, Incorporated (26). This polymer is designed for use in a variety of disposable products, such as trash bags, food service ware, and packaging. When used in films and sheets, Mater-Bi biodegrades under composting conditions in 20–45 days. Biodegradation for injection-molded items typically takes 75–120 days. The compost acts as a soil amendment for agricultural and horticultural use.

Principle 11: Analytical methodologies need to be further developed to allow for real-time, in-process monitoring and control prior to the formation of hazardous substances

Pollution prevention can be facilitated through on-line monitoring. Process analytical chemistry can be used to detect the generation of hazardous by-products, permitting reaction conditions to be adjusted to minimize or eliminate the formation of these unwanted substances. In-process monitoring may also be employed to follow the progress of a reaction, reducing the need for excess reagents.

Georgakis and co-workers (27) have used on-line monitoring to study semibatch emulsion polymerization of styrene. Conversion is monitored via an on-line digital densitometer, providing near-continuous feedback. A variation of capillary hydrodynamic fractionation (CHDF) is used for on-line particle size measurements. The data obtained from the on-line sensors were used to support a dynamic growth model but could also be used to direct the process evolution.

Principle 12: Substances and the form of a substance used in a chemical process should be chosen so as to minimize the potential for chemical accidents, including releases, explosions, and fires

Toxicity and pollution are not the only concern with regard to chemicals: flammability, volatility, and shock sensitivity present potential problems as well. Because accidents do happen, it is important to use the safest form of a chemical possible to minimize harm in the event of an accident. At times, pollution prevention measures may increase accident potential, as in the case of recycling a highly flammable solvent. Compromise may be necessary to realize the most environmentally sound process available.

Nowhere is the safety of materials used more important than in the interior of an airplane. Polymers are ubiquitous in airplanes. When a fire breaks out in an airplane, lives may be lost as the result of inhalation of toxic gases released by combustion of polymers. An ideal material would resist decomposition and avoid the release of toxic and combustible gases upon burning. Westmoreland (28) has identified polyhydroxyamide (PHA) as a fire-resistant thermoplastic that can be molded into seats, overhead bins, and other airplane parts. When heated, PHA degrades into water and polybenzoxazole (PBO), a hard, inert polymer (Fig. 2.11). The use of safer materials in airplanes may permit a higher survival rate in the event of a plane crash and fire.

Although widely used in a variety of applications, PVC (polyvinyl chloride) presents environmental concerns. PVC contains phthalates and chlorine, which are possible endocrine disruptors, and generates dioxin by-products during manufacture and combustion. An alternative to PVC is provided by metallocene polyolefins (29). These polymers demonstrate superior physical properties to PVC and are man-

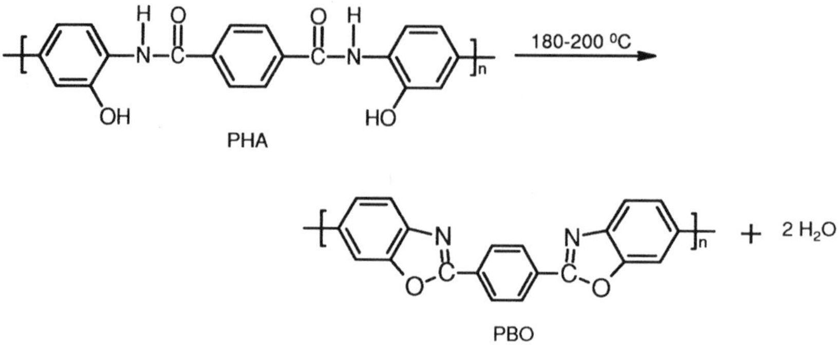

Figure 2.11   Thermal degradation of PHA to PBO and water.

ufactured without toxic plasticizers. The metallocene catalysts allow the polymer properties to be tailored during the polymerization process. Potential applications for this new technology include flooring, siding, and automotive finishes.

2.3   DESIGNING SAFER CHEMICALS

When designing a reaction, chemists need to evaluate all aspects of the process: Are the reagents used or products isolated toxic? Is an organic solvent necessary? Does the reaction require heating or cooling? How will products be separated? Is a protecting group needed? Although it is tempting to stick with familiar reactions, rather than search for benign alternatives, gains in pollution prevention will remain elusive if chemists are satisfied with the status quo.

Safer chemicals are those that are less toxic than those in current use or those that are safer to handle (lower accident potential). The substitution of dimethylcarbonate for phosgene in carbonylation reactions provides an excellent example of using a less toxic substance to synthesize the same product (30). Recently, the conversion of a ketone to a lactone has been accomplished using genetically engineered baker's yeast (31), a safer alternative to the shock-sensitive and explosive m-chloroperoxybenzoic acid commonly in use.

Developing safer chemicals draws on our current understanding of functional groups and structural features that may pose a threat to the environment or human health. A safer chemical is one that does not include functional groups that are known to be toxic to living organisms. A chemical that breaks down into harmless degradation products at the end of its useful lifetime is better than one that is persistent in nature. A safer chemical presents a reduced risk of fire or explosion, thereby decreasing the potential for accidents. By evaluating the hazard potential for reactants and products, chemists can make informed decisions regarding safer alternatives. A safer chemical will perform the same function as a chemical currently in use but will do so in a more benign manner.

REFERENCES

1. Anastas, P. T., and J. C. Warner. 1998. *Green Chemistry: Theory and Practice.* New York: Oxford University.

2. *The Pollution Prevention Act.* 1990. 42 U.S.C. §§13101–13109.

3. Browner, C. M. *EPA Journal* 1993. 19(3): 6–8.

4. *The Emergency Planning and Community Right-to-Know Act.* 1986. 42 U.S.C. §§11001–11050.

5. U.S. Environmental Protection Agency. Office of Pollution Prevention and Toxics. 1997. *The Presidential Green Chemistry Challenge Awards Program: Summary of 1997 Award Entries and Recipients,* p. 33. EPA744-S-97-001. Washington, D.C.

6. U.S. Environmental Protection Agency. Office of Pollution Prevention and Toxics. 1997. *The Presidential Green Chemistry Challenge Awards Program: Summary of 1997 Award Entries and Recipients,* p. 30. EPA744-S-97-001. Washington, D.C.

7. Trost, B. M. *Science* 1991. 254: 1471–1477.

8. Sheldon, R. A. *Chemtech* 1994. March: 38–47.

9. *Science* 1999. 284: 243.

10. Tsuda, T., K. Maruta, and Y. Kitaike. *J. Am. Chem. Soc.* 1992. 114: 1498–1499.

11. U.S. Environmental Protection Agency. Office of Pollution Prevention and Toxics. 1997. *The Presidential Green Chemistry Challenge Awards Program: Summary of 1997 Award Entries and Recipients,* p. 36. EPA744-S-97-001. Washington, D.C.

12. U.S. Environmental Protection Agency. Office of Pollution Prevention and Toxics. 1998. *The Presidential Green Chemistry Challenge Awards*

Program: Summary of 1998 Award Entries and Recipients, p. 64. EPA744-R-98-001. Washington, D.C.

13. U.S. Environmental Protection Agency. Office of Pollution Prevention and Toxics. 1998. *The Presidential Green Chemistry Challenge Awards Program: Summary of 1998 Award Entries and Recipients,* p. 17. EPA744-R-98-001. Washington, D.C.

14. Komiya, K., S. Fukuoka, M. Aminaka, K. Hasegawa, H. Hachiya, H. Okamoto, T. Watanabe, H. Yoneda, I. Fukawa, and T. Dozono. 1996. In *Green Chemistry: Designing Chemistry for the Environment,* ed. P. T. Anastas and T. C. Williamson. ACS Symposium Series 626. American Chemical Society: Washington, D.C., p. 20–32.

15. Buelow, S., P. Dell'Orco, D. Morita, D. Pesiri, E. Birnbaum, S. Borkowsky, G. Brown, S. Feng, L. Luan, D. Morgenstern, and W. Tumas. 1998. In *Green Chemistry: Frontiers in Benign Chemical Syntheses and Processes,* ed. P. T. Anastas and T. C. Williamson. New York: Oxford University Press, p. 265–285.

16. Mistele, C. D., and J. M. DeSimone. 1998. In *Green Chemistry: Frontiers in Benign Chemical Syntheses and Processes,* ed. P. T. Anastas and T. C. Williamson. New York: Oxford University Press, p. 286–311.

17. DeSimone, J. M., Z. Guan, and C. S. Elsbernd. 1992. *Science* 257: 945–947.

18. U.S. Environmental Protection Agency. Office of Pollution Prevention and Toxics. 1996. *The Presidential Green Chemistry Challenge Awards Program: Summary of 1996 Award Entries and Recipients,* p. 35–36. EPA744-K-96-001. Washington, D.C.

19. U.S. Environmental Protection Agency. Office of Pollution Prevention and Toxics. 1996. *The Presidential Green Chemistry Challenge Awards Program: Summary of 1996 Award Entries and Recipients,* p. 5. EPA744-K-96-001. Washington, D.C.

20. U.S. Environmental Protection Agency. Office of Pollution Prevention and Toxics. 1998. *The Presidential Green Chemistry Challenge Awards Program: Summary of 1998 Award Entries and Recipients,* p. 9. EPA744-R-98-001. Washington, D.C.

21. U.S. Environmental Protection Agency. Office of Pollution Prevention and Toxics. 1997. *The Presidential Green Chemistry Challenge Awards Program: Summary of 1997 Award Entries and Recipients,* p. 32. EPA744-S-97-001. Washington, D.C.

22. Bisht, K. S., L. A. Henderson, Y. Y. Svirkin, R. A. Gross, D. L. Kaplan, and G. Swift. 1998. In *Enzymes in Polymer Synthesis,* ed. R. A. Gross,

D. L. Kaplan, and G. Swift. ACS Symposium Series. American Chemical Society: Washington, D.C., p. 90–111.

23. *Chem. Eng. News* 1998. 76 (27): 11–16.

24. Swift, G. 1998. *Green Chemistry and Engineering Conference Proceedings.* 2nd Annual Green Chemistry and Engineering Conference, Washington, D.C.

25. U.S. Environmental Protection Agency. Office of Pollution Prevention and Toxics. 1997. *The Presidential Green Chemistry Challenge Awards Program: Summary of 1997 Award Entries and Recipients*, p. 20. EPA744-S-97-001. Washington, D.C.

26. U.S. Environmental Protection Agency. Office of Pollution Prevention and Toxics. 1998. *The Presidential Green Chemistry Challenge Awards Program: Summary of 1998 Award Entries and Recipients*, p. 27. EPA744-R-98-001. Washington, D.C.

27. Liotta, E. D. Sudol, M. S. El-Aasser, and C. Georgakis. 1998. *J. Polym. Sci. Polym. Chem. Ed.* 36: 1553–1571.

28. Westmoreland, P. R. 1999. *Chem. Eng. News* 77(17): 26.

29. U.S. Environmental Protection Agency. Office of Pollution Prevention and Toxics. 1998. *The Presidential Green Chemistry Challenge Awards Program: Summary of 1998 Award Entries and Recipients*, p. 59–60. EPA744-R-98-001. Washington, D.C.

30. Rivetti, F., U. Romano, and D. Delledonne. 1996. In *Green Chemistry: Designing Chemistry for the Environment,* ed. P. T. Anastas and T. C. Williamson. ACS Symposium Series 626. American Chemical Society: Washington, D.C., p. 70–80.

31. U.S. Environmental Protection Agency. Office of Pollution Prevention and Toxics. 1998. *The Presidential Green Chemistry Challenge Awards Program: Summary of 1998 Award Entries and Recipients*, p. 12–13. EPA744-R-98-001. Washington, D.C.

CHAPTER 3

REGULATION OF POLYMERS UNDER THE TOXIC SUBSTANCES CONTROL ACT

3.1 BACKGROUND

Passage of the Toxic Substances Control Act (TSCA) in 1976 gave the Environmental Protection Agency (EPA) the authority to regulate new and existing chemicals in order to protect human health and the environment (1). TSCA allows the Agency to limit the production, use, or disposal of a new chemical if it presents an unreasonable risk to the environment. In addition, the EPA is authorized to require testing of new chemicals. The foundation for TSCA legislation was a 1971 report by the President's Council on Environmental Quality (CEQ) (2). This report concluded that potentially damaging toxic substances were entering the environment and that there were no laws in existence to regulate the manufacture, importation, or use of toxic chemicals in the United States. Consequently, the CEQ proposed a new program to test for and control toxic substances, a recommendation that promulgated TSCA legislation.

TSCA regulates organic and inorganic chemical substances and mixtures. Food and food additives, cosmetics, drugs, nuclear materials, pesticides, and tobacco are addressed by separate legislation, such as the Federal Food, Drug, and Cosmetic Act (3) and the Federal Insecticide, Fungicide, and Rodenticide Act (4). Under TSCA, the Agency has the authority to

(1)   require manufacturers and importers to submit information on all new chemical substances prior to manufacture for commercial purposes;

(2)   require manufacturers and processors to collect, maintain, and possibly submit information on chemical substances; and

(3)   regulate new and existing chemical substances that are expected to present or are presenting unreasonable risks to health and the environment.

The goal of TSCA is to assess the potential risks of a chemical *before* it is commercially manufactured rather than after a new chemical has been introduced into commerce. The EPA is required to balance the risk of using a chemical substance against the benefits gained from its use, without hindering technological innovation.

The TSCA Chemical Substance Inventory defines those chemical substances in existence in United States commerce for purposes of implementing TSCA but is not a list of toxic chemicals. The Inventory currently lists more than 75,000 chemical substances whose manufacture or processing has taken place since January 1, 1975. It is the responsibility of manufacturers and importers to determine if a chemical substance is listed on the Inventory.

3.2   NEW CHEMICAL RESPONSIBILITIES

The manufacture and importation of new chemical substances are regulated under section 5 of TSCA. Specifically, section 5(a)(1) of TSCA requires that the EPA be notified at least 90 days before a new chemical substance is manufactured or imported for commercial purposes. A "new chemical substance" is any substance not on the Inventory of Chemical Substances compiled by the EPA under section 8(b) of TSCA. Submitters must also file a Significant New Use Notice (SNUN) when intending to use a chemical substance already on the Inventory but in a manner that differs substantively from that already approved. The potential hazard of a new chemical substance is evaluated through the Premanufacture Notification (PMN) process (5).

When filing a PMN, manufacturers must provide the following information:

- the common or trade name, chemical identity, and molecular structure of each chemical substance;
- the categories or proposed categories of use for each substance;
- estimates of the total amounts of substances that will be manufactured and used, broken out by category of use;
- a description of by-products generated during the manufacturing process; and
- an estimation of the number of individuals exposed, and the exposure duration, and the method of disposal of the substance.

Manufacturers are not required to submit test data for a new chemical substance when submitting a PMN but are asked to provide any available test data, both internal and external, related to effects on health or the environment. Often, little information may be available on the health and environmental effects of a new chemical. Consequently, the EPA uses a variety of tools in assessing the potential harm that a new substance may pose to human health and the environment.

Four steps comprise the PMN Review Process: chemistry review, hazard (toxicity) evaluation, exposure evaluation, and risk assessment/risk management (6). Low-risk substances are dropped early in the process so that limited resources may be focused on those substances that pose a greater risk to human health and the environment. Although a new chemical substance may be dropped from a full review, manufacture of the substance may not begin before the normal 90-day review period has expired.

The initial chemistry review ascertains the completeness and accuracy of the submission. Exemption notices are reviewed for compliance with the exemption guidelines. An Inventory review verifies the accuracy of the chemical name of the new substance and determines if the substance is already on the Inventory. The Chemistry Report summarizes the chemistry of the substance, including synthesis, presence of by-products and impurities, uses, production volumes, and manufacturing processes. Because most PMN submissions contain few physicochemical data, these properties are often estimated from structural analogs. Estimates are made to maximize exposure and hazard potential, thereby providing a margin of safety.

The Chemistry Report provides the basis for the hazard evaluation. An interdisciplinary team of scientists identifies the potential hazards of the PMN substance to human health and the environment. Analysis of hazardous properties considers the potential of absorption and metabolism in humans, human toxicity, toxicity to environmental organisms, and environmental fate. The Agency relies on structure-activity relationships (SARs) to estimate toxicity, due to the lack of toxicity data available for most PMN substances. With SARs, the toxicity of a new chemical substance is approximated by comparison with structurally similar compounds for which toxicological and environmental data are known. Quantitative structure-activity relationships (QSARs) provide a more precise estimate of toxicity by examining physicochemical properties in conjunction with structure-activity relationships. Predicting the environmental fate of a chemical plays a key role in hazard identification by considering relative rates of biodegradation, hydrolysis, and photolysis; adsorption to soils and sediments; treatability; and half-lives in the atmosphere, soils, sediments, and surface waters. Properties used to predict the environmental fate of a new chemical include water solubility, octanol/water partition coefficient, and vapor pressure.

The exposure evaluation phase of the PMN review identifies the likelihood and magnitude of exposure of a substance to workers, consumers, the general population, and the environment. Physicochemical properties play a pivotal role in estimating exposure. Vapor pressure and molecular weight, for example, are important in determining inhalation and dermal exposure. Other factors considered during the exposure evaluation process include expected environmental releases, commercial and consumer use, and environmental fate.

The final step in the PMN review process is the risk assessment/ risk management phase. Risk assessment characterizes the potential for harm to human health and the environment from exposure to a chemical substance. Risk management identifies ways in which the risks posed by a chemical substance can be minimized. Policy alternatives are evaluated, along with regulatory and nonregulatory actions. Risks and benefits are balanced in making risk management decisions about the PMN substance. Approximately 90% of the substances submitted for PMN review complete the process without the imposition of restrictions or regulations.

3.3 AUTHORITY FOR EXEMPTIONS

Section 5(a)(1) of TSCA requires manufacturers and importers of new chemical substances, used for commercial purposes, to notify the EPA at least 90 days before beginning the manufacturing process. A "new chemical substance" is any substance not listed on the Inventory of Chemical Substances compiled by EPA under section 8(b) of TSCA. EPA is authorized under section 5(h)(4) of TSCA to exempt the manufacturer or importer of any new chemical substance from part or all of the provisions of section 5 if the Agency determines that no unreasonable risk to human health or the environment is presented by the manufacturing, processing, distribution, use, or disposal of the new chemical substance. Exemptions from PMN requirements may be granted in specific cases, as outlined below.

(1) New chemical substances used solely for test marketing purposes may be granted an exemption, provided there is no unreasonable risk to human health or the environment. The exemption allows a company to evaluate the commercial viability of a new chemical and receive customer feedback before filing a PMN. The manufacturer must submit a test market exemption application (TMEA) to the EPA. Regulations governing TMEAs may be found in the PMN rule and instruction manual (5, 6), and the New Chemical Information Bulletin (7). TMEAs require no user fee and are reviewed by the Agency within 45 days.

(2) Chemical substances that are manufactured in small quantities exclusively for scientific experimentation, analysis, or research and development are exempt from PMN reporting. Additional information on this exemption may be found in two Federal Register notices (8, 9) and a *New Chemical Information Bulletin* (7).

(3) A partial exemption from TSCA section 5 reporting requirements is granted to manufacturers of chemical substances produced in quantities less than 1,000 kg per 12-month period (10). Manufacturers must submit a low-volume exemption notice to the EPA 30 days before commencement of manufacture.

(4)   Substances may qualify for a low release and exposure exemption (loREX) if they meet the release and exposure criteria stated in the rule (11). Manufacturers must submit an exemption notification at least 30 days before beginning production.

(5)   Manufacturers of instant film may commence manufacture of new chemical substances used in instant photographic articles immediately after submitting an exemption notice to EPA. Distribution of the new chemical in commerce may not begin until a PMN has been submitted and the review period has expired. Exposure must be contained until PMN review is complete (12).

(6)   Manufacturers of polymers that meet certain criteria and are deemed to pose no unreasonable threat to human health and the environment may be granted an exemption from PMN reporting. Details of this exemption form the basis of this book.

The original polymer exemption rule was promulgated in 1984 (13). The final polymer exemption rule (14) reflects the Agency's extensive experience in reviewing polymers submitted under the PMN process and the 1984 polymer exemption rule. As a result of evaluating thousands of polymers, EPA has developed guidelines to identify polymers that do not present an unreasonable risk of harm to human health and the environment. Several factors contributed to the refinement of these guidelines: review of the available literature on the toxicity of polymers, analyses of portions of the PMN polymer data base, information provided to EPA by external groups, and professional judgment of EPA staff scientists. The final polymer exemption rule allows more low-risk polymers to qualify for exemption while excluding those polymers that require review under the 90-day PMN process.

The polymer exemption rule eases the reporting requirements for industry. Manufacturers and importers of polymers that meet the polymer exemption criteria no longer need to submit an exemption application before commencement of manufacturing. Companies must now submit an annual report on exempt polymers for which manufacture or importation began during the previous calendar year. Records that document a polymer's eligibility for exemption must be maintained at the manufacturing site. Critical data that are also required include the date on which manufacture began, production volume for the first 3 years, and chemical structure and identity. The polymer exemption rule is designed to reduce industry's regulatory reporting responsibilities, thereby promoting the manufacture of safer polymers.

3.4 OUTLINE OF THE APPROACH

In 1990, Congress passed the Pollution Prevention Act (15). This act established pollution prevention as the EPA's top priority in addressing environmental problems. Historically, waste has been treated as a problem to be dealt with at the "end-of-the-pipe." Pollution prevention advocates eliminating or minimizing waste, thereby avoiding the costs and technological challenges associated with waste treatment.

As a class, polymers are relatively unreactive. They are not readily transported across biological membranes at molecular weights greater than or equal to 10,000. As stated in the Agency's 1984 polymer exemption rule

for a chemical to elicit a toxic response within an organism, it must come into direct contact with the biological cells from which it elicits the response. Because all organisms are encased in protective membranes, a chemical must usually penetrate these membranes and be translocated to various parts of the organism to gain access to target sites. Therefore, it can be reasoned that if a chemical cannot penetrate the protective membranes to gain access to a target site, it usually cannot elicit a response in the organism no matter what inherent potential it may have to do so. It can be further reasoned that if a chemical cannot elicit a response, it will generally not present a risk. (13)

In accord with the final rule, polymers meeting the exemption criteria are of low concern due to their lack of reactivity and their molecular size. The Agency's risk analysis of polymers is based on

- the effect of molecular weight on the overall risk,
- the specific concerns presented by polymers submitted as PMNs,
- toxicological data on particular polymers,

- an analysis of the Agency's database on over 500 polymers to determine the potential for exposure to respirable particles of insoluble high molecular weight polymers, and
- an analysis of the Agency's experience in assessing potential risks of new polymers submitted as PMNs or exemption applications.

The final rule excludes from exemption those classes of polymers that pose a potential threat to human health or the environment. Examples of such substances include polymers that are expected to substantially degrade, decompose, or depolymerize; certain water-absorbing polymers; and polymers that contain specific reactive functional groups. Limits on oligomer content are also specified, as the lower molecular weight oligomers are more reactive than the resulting polymers.

A polymer that meets the exemption criteria is expected to pose no unreasonable risk of harm to human health and the environment. Therefore, a polymer that is designed to meet the criteria outlined in the polymer exemption rule is inherently a safer polymer. Designing safer polymers provides economic benefits to both industry and government. Industry saves on user fees and reporting and delay costs due to the elimination of submission requirements. The expanded criteria also provide industry with increased flexibility, as more polymers are eligible for the exemption. Fewer submissions allow the EPA to focus its resources on substances that present unknown or potentially higher risks to human health and the environment.

REFERENCES

1. The Toxic Substances Control Act (TSCA). 1976. 15 U.S.C. §§2601–2629 (1982 and Supp. III 1985).
2. Council on Environmental Quality (CEQ). 1976. *Toxic Substances. Legislative History of the Toxic Substances Control Act (TSCA Legislative History)*. Reprinted in Staff House Committee on Interstate and Foreign Commerce, 94th Cong. 2nd sess., 1971 (April). New York: U.S. Government Printing Office, December, p. 757–788.
3. The Federal Food, Drug, and Cosmetic Act (FFDCA). 1982. 21 U.S.C. §§301–392.

4. The Federal Insecticide, Fungicide, and Rodenticide Act (FIFRA). 7 U.S.C. §136 et seq.

5. U.S. Environmental Protection Agency (USEPA). Office of Toxic Substances. 1986. *New Chemical Review Process Manual.* Washington, D.C. EPA report number: EPA-560/3-86-002.

6. For a detailed description of the PMN review process, see DeVito, S. C., and C. A. Farris. 1997. *Premanufacture Notification: Chemistry Assistance for Submitters.* New York: Wiley-Interscience.

7. U.S. Environmental Protection Agency. *New Chemical Information Bulletin. Exemptions for Research and Development and Test Marketing.* Washington, D.C.

8. U.S. Environmental Protection Agency. 1984 (December 27). Premanufacture Notification. Proposed Revisions of Regulation (49 FR 50202).

9. U.S. Environmental Protection Agency. 1986 (April 22). Toxic Substances; Revisions of Premanufacture Notification Regulations; Final Rule. 40 CFR Part 720 (51 FR 15096).

10. U.S. Environmental Protection Agency. 1985 (April 26). Premanufacture Notification Exemption; Exemption for Chemical Substances Manufactured in Quantities of 1,000 kg or Less Per Year. 40 CFR Part 723 (50 FR 16477).

11. U.S. Environmental Protection Agency. 1995 (March 29). Premanufacture Notification Exemption; Revision of Exemption for Chemical Substances Manufactured in Small Quantities; Low Release and Exposure Exemption; Final Rule. 40 CFR Part 723 (60 FR 16336–16351).

12. U.S. Environmental Protection Agency. 1982 (June 4). Premanufacture Notification; Exemption for Chemicals Used in or for the Manufacture or Processing of Instant Photographic and Peel-Apart Film Articles (47 FR 24308). Also in 40 CFR 723.175.

13. U.S. Environmental Protection Agency. 1984 (November 21). Premanufacture Notification Exemptions; Exemptions for Polymers; Final Rule (49 FR 46066). Also in 40 CFR 723.

14. U.S. Environmental Protection Agency. 1995 (March 29). Premanufacture Notification Exemptions; Revisions of Exemptions for Polymers; Final Rule. 40 CFR Part 723 (60 FR 16316–16336).

15. The Pollution Prevention Act (PPA). 1990. 42 U.S.C. §§ 13101–13109.

CHAPTER 4

THE TSCA POLYMER EXEMPTION: PROVISIONS AND GUIDANCE

4.1 BACKGROUND

To design a safer polymer, it is important to first understand the nature of the hazards that one is striving to avoid. The Polymer Exemption Rule was developed as the outcome of a systematic review of over 20,000 polymers by the United States Environmental Protection Agency (EPA). Through this review, the EPA was able to identify and characterize those properties that indicate that a polymer may be of concern to human health or the environment. [Note: It is important that the reader understand that not all of the criteria are direct indications of hazard. Restrictions on biodegradability, for example, are included because the uncertainty of the resulting degradation products was thought by the EPA to necessitate regulatory review.] By understanding which properties, therefore, can cause potential hazard to human health and the environment, the designer of a new polymer can, through structural and molecular manipulation, often eliminate or reduce those hazards from being present in the polymeric product.

The information contained within this chapter serves a dual purpose. First, the guidelines presented assist the regulated community in determining if a new chemical substance is eligible for the polymer exemption. Second, by designing a polymer to meet the exemption criteria, the manufacturer is taking the initiative in designing a safer polymer product. Although the information presented here provides guidance in determining the eligibility of a polymer for exemption, it is not intended to substitute for or supersede the regulations found in 40 CFR §723.250 (1) and the Federal Register (USEPA 1995; 2). These documents are essential in determining compliance with all procedural and record keeping requirements of the polymer exemption.

4.2   DEFINITIONS

The criteria used to determine the eligibility of a polymer for exemption are based on a shared understanding of basic terminology. Table 4.1 lists the definitions used in the implementation of the polymer exemption rule. The polymer definition itself conforms with that adopted by the Organization for Economic Co-operation and Development (OECD) countries in May 1993 (3).

4.3   ELIGIBILITY REQUIREMENTS

The following requirements must be met in order for a new chemical substance to be eligible for exemption under the amended rule:

- The substance must meet the definition of a polymer [see 40 CFR §723.250(b)];
- The substance must not be specifically excluded from the polymer exemption [see 40 CFR §723.250(d)]; and
- The substance must meet one of the exemption criteria [see 40 CFR §723.250(e)(1), (e)(2), or (e)(3)].

4.3.1   Meeting the Definition of a Polymer

A chemical substance must meet the definition of a polymer to be eligible for the polymer exemption rule. A polymer satisfies the sequence and distribution criteria if

TABLE 4.1. Definitions Applicable to the Polymer Exemption Rule

Polymer: a chemical substance consisting of molecules characterized by the sequence of 1 or more types of monomer units and comprising a simple weight majority of molecules containing at least 3 monomer units that are covalently bound to at least 1 other monomer unit or other reactant and that consists of less than a simple weight majority of molecules of the same molecular weight. Such molecules must be distributed over a range of molecular weights, wherein differences in the molecular weight are primarily attributable to differences in the number of monomer units.

Monomer: a chemical substance that is capable of forming covalent bonds with 2 or more like or unlike molecules under the conditions of the relevant polymer-forming reaction used for the particular process.

Monomer unit: the reacted form of the monomer in the polymer.

Sequence: a continuous string of monomer units within the molecule that are covalently bonded to one another and are uninterrupted by units other than monomer units.

Reactant: a chemical substance that is used intentionally in the manufacture of a polymer to become chemically a part of the polymer composition. (Reactants include monomers, chain transfer and cross-linking agents, monofunctional groups that act as modifiers, other end groups or pendant groups incorporated into the polymer. For example, sodium hydroxide is considered a reactant when the sodium ion becomes part of the polymer molecule as a counter ion.)

Other reactant: a molecule linked to 1 or more sequences of monomer units but which under the relevant reaction conditions used for the particular process cannot become a repeating unit in the polymer structure. (This term is used primarily in applying the concept of sequence in the definition of a polymer.)

Polymer molecule: a molecule that contains a sequence of at least 3 monomer units, which are covalently bound to at least 1 other monomer unit or other reactant.

Internal monomer unit: a monomer unit of a polymer molecule that is covalently bonded to at least 2 other molecules. Internal monomer units of polymer molecules are chemically derived from monomer molecules that have formed covalent bonds between 2 or more other monomer molecules or other reactants.

TABLE 4.1. (*Continued*)

Number-average molecular weight (M_n): the arithmetic average (mean) of the molecular weights of all molecules in a polymer. (This value should not take into account unreacted monomers and other reactants but must include oligomers.)

Dalton: precisely 1.0000 atomic mass unit or 1/12 the mass of a carbon atom of mass 12. Hence, a polymer with a molecular weight of 10,000 atomic mass units has a mass of 10,000 daltons.

Oligomer: a low molecular weight species derived from the polymerization reaction. The Organization for Economic Cooperation and Development (OECD) has drafted guidelines for determining the low molecular weight polymer content (3).

- greater than 50% of the molecules are composed of a sequence of at least three monomer units plus at least one additional monomer unit or other reactant; that is, >50% of the substance is composed of polymer molecules; and
- the amount of polymer molecules of any one molecular weight does not exceed 50% by weight.

The following examples illustrate the application of the polymer definition criteria to specific chemical substances. Some of the examples have been taken from the Chairman's Report (4) of the Chemicals Group and Management Committee at the Third Meeting of OECD Experts on Polymers (Tokyo, April 14–16, 1993) in which the Agency participated. Examples 1–5 demonstrate the sequence criteria for defining a polymer molecule. The term "o.r." refers to other reactant and the term "m.u." refers to monomer unit.

Example 1

Ethoxylation of benzenetetrol with ethylene oxide yields the structure shown in Figure 4.1. The compound shown does not meet the sequence criterion and, therefore, is not a polymer molecule because there is no sequence of three monomer units (m.u.) from ethylene oxide. Benzenetetrol is classified as an other reactant (o.r.) because the phenol hydroxy group cannot react with another phenol hydroxy or with an opened epoxide under the reaction conditions.

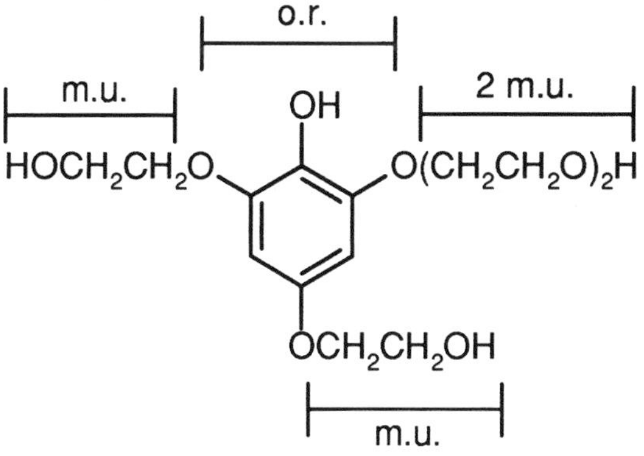

Figure 4.1 Ethoxylated benzenetetrol (Example 1).

Example 2

The structure shown in Figure 4.2 fits the definition of a polymer molecule if $n \geq 3$. The compound is the result of ethoxylation of hydroquinone with ethylene oxide. Hydroquinone is considered an other reactant (o.r.) because the phenol hydroxy group cannot react with another phenol hydroxy or with an opened epoxide under the reaction conditions.

Example 3

Ethoxylation of glycerol with ethylene oxide produces the structure shown in Figure 4.3. This compound qualifies as a polymer molecule if at least seven equivalents of ethylene oxide are charged to the reaction vessel. Each hydroxyl may be ethoxylated twice or less if

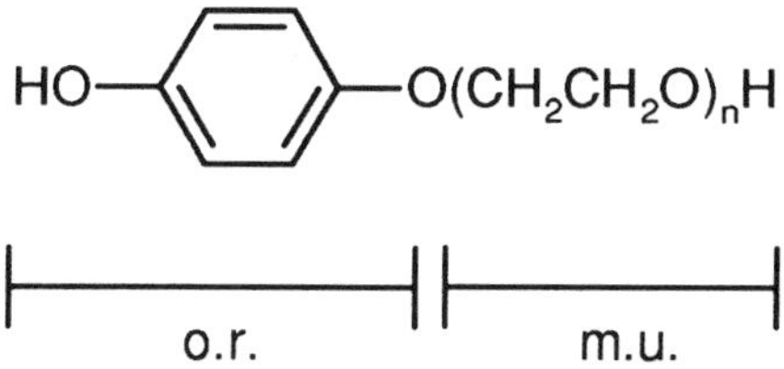

Figure 4.2 Ethoxylated hydroquinone (Example 2).

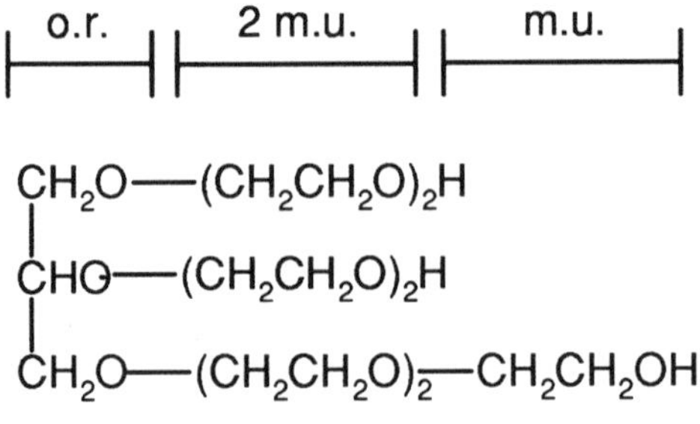

$$\underset{\vdash\!\!\!\!-\!\!\!\!-\!\!\!\!\dashv}{\text{o.r.}}\underset{\vdash\!\!\!\!-\!\!\!\!-\!\!\!\!-\!\!\!\!-\!\!\!\!\dashv}{\text{2 m.u.}}\underset{\vdash\!\!\!\!-\!\!\!\!-\!\!\!\!-\!\!\!\!\dashv}{\text{m.u.}}$$

$$CH_2O\text{---}(CH_2CH_2O)_2H$$
$$CHO\text{---}(CH_2CH_2O)_2H$$
$$CH_2O\text{---}(CH_2CH_2O)_{\overline{2}}\text{---}CH_2CH_2OH$$

Figure 4.3   Ethoxylated glycerol (Example 3).

fewer than seven equivalents are used, which would not satisfy the sequence criterion for polymers.

Example 4

The structure shown in Figure 4.4 is not a polymer molecule because there are no repeating units. Both the glycerol and the fatty acid are classified as other reactants because they cannot repeat under the reaction conditions. Methylene (CH_2) is not a monomer unit because it remains unchanged in both the reactant and the product. Methylene would have to be the reacted form of the monomer present in the polymer in order to qualify as a monomer unit.

Example 5

The epoxy resin in Figure 4.5 qualifies as a polymer molecule because it contains a sequence of three monomer units and one other reactant.

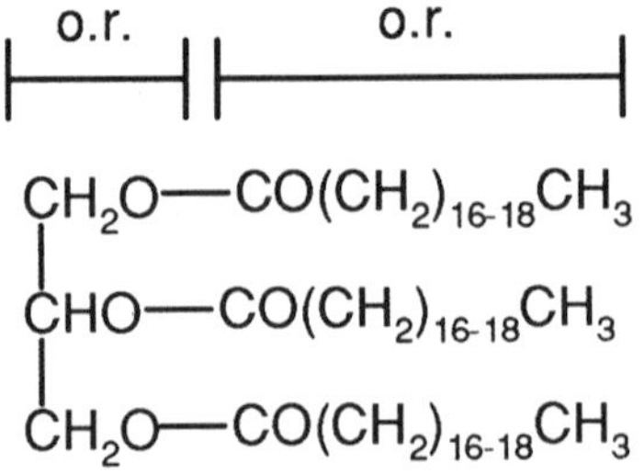

$$\underset{\vdash\!\!\!\!-\!\!\!\!-\!\!\!\!\dashv}{\text{o.r.}}\underset{\vdash\!\!\!\!-\!\!\!\!-\!\!\!\!-\!\!\!\!-\!\!\!\!-\!\!\!\!\dashv}{\text{o.r.}}$$

$$CH_2O\text{---}CO(CH_2)_{16\text{-}18}CH_3$$
$$CHO\text{---}CO(CH_2)_{16\text{-}18}CH_3$$
$$CH_2O\text{---}CO(CH_2)_{16\text{-}18}CH_3$$

Figure 4.4   Glycerol triester (Example 4).

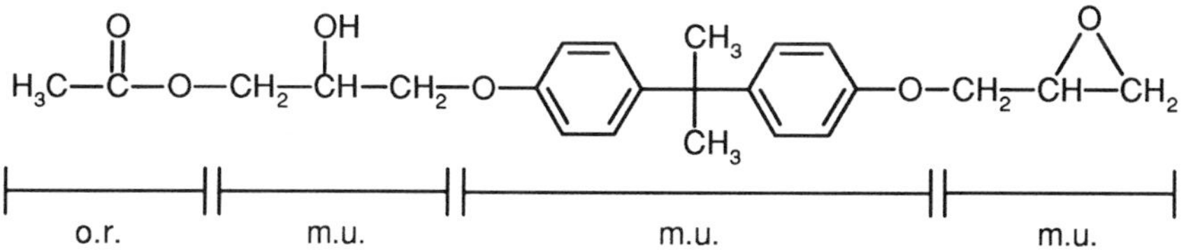

Figure 4.5   Epoxy resin (Example 5).

Examples 6, 7, and 8

The data in Table 4.2 illustrate the application of the sequence and distribution rules to the reaction between ethylene oxide (EO, monomer unit) and an other reactant derived from an alcohol (RO). Example 6 meets the polymer definition because more than 50% of the products have molecules containing three or more monomer units plus one monomer unit or other reactant, yet no single species is present in a concentration >50%. Example 7 would not be classified as a polymer because <50% of the products consists of three monomer units plus one monomer unit or other reactant. Example 8 does not satisfy the polymer definition because one species is present at a concentration >50%.

Example 9

The enzyme pepsin meets the sequence requirements for a polymer but not the distribution requirement. The molecular weight of pepsin is identical for all molecules; therefore, one species is present in >50%, and pepsin may not be listed as a polymer. In contrast, a

TABLE 4.2. Distribution Criteria for Ethoxylated Alcohols (Examples 6, 7, and 8)

Species	o.r. + m.u.	Example 6	Example 7	Example 8
RO.EO.H	1 + 1	5%	25%	8%
RO.EO.EO.H	1 + 2	20%	35%	20%
RO.EO.EO.EO.H	1 + 3	30%	20%	52%
RO.EO.EO.EO.EO.H	1 + 4	40%	10%	10%
RO.EO.EO.EO.EO.EO.H	1 + 5	5%	10%	10%

lipoprotein or mucoprotein may qualify as a polymer if the lipo- and muco-portions vary sufficiently such that no single species is present at a concentration >50%.

4.4 SUBSTANCES EXCLUDED FROM THE EXEMPTION

Certain categories of polymers are excluded from exemption under the new polymer exemption rule because the Agency cannot determine if these substances pose an unreasonable risk of injury to human health or the environment. The history behind the selection of these categories may be found in the preamble to the 1995 polymer exemption rule (USEPA 1995; 2). The excluded categories are discussed in the following sections.

4.4.1 Exclusions for Cationic and Potentially Cationic Polymers

Cationic polymers and polymers that are likely to become cationic in an aqueous environment are ineligible for the polymer exemption rule and may not be manufactured under it. The likelihood of a polymer becoming cationic is based on such factors as the nature of the polymer precursors, the type of reaction, the type of manufacturing process, the products generated during polymerization, the intended uses of the substance, and the conditions associated with use of the substance. Toxicity toward aquatic organisms is the primary concern for this class of polymers.

> **Cationic polymer:** a polymer that contains a net positively charged atom(s) or associated group(s) of atoms covalently linked to the polymer molecule, such as phosphonium, sulfonium, and ammonium cations.

> **Potentially cationic polymer**: a polymer containing groups that are reasonably anticipated to become cationic, such as amines (primary, secondary, tertiary, aromatic, etc.) and all isocyanates (which hydrolyze to form carbamic acids, followed by decarboxylation to form amines).

4.4.2   Cationic Polymers Not Excluded from Exemption

On the basis of a review of thousands of polymers, the Agency has identified two classes of cationic and potentially cationic polymers that do not present an unreasonable risk of injury to human health or the environment. These two classes may be eligible for the exemption and are as follows:

(a)   Cationic or potentially cationic polymers that are solids, are only to be used in the solid phase, are neither water soluble nor dispersible in water, and are not excluded from exemption by other factors.

(b)   Cationic or potentially cationic polymers with low cationic density (percentage of cationic or potentially cationic species with respect to the overall weight of the polymer) that are not excluded from exemption by other factors. A functional group equivalent weight $\geq 35,000$ daltons qualifies a polymer as having low cationic density.

Functional group equivalent weight (FGEW): the weight of polymer that contains one equivalent of the functional group or the ratio of the number-average molecular weight (NAVG MW) to the number of functional groups in the polymer. Methods for calculating FGEW are presented in Section 4.9.

Example 10

Combining equal molar amounts of ethanediamine and phthalic acid yields a polymer with an equal number of unreacted amine and carboxylic acid groups. On average, each polymer molecule would contain one potentially cationic amine group and one unreacted carboxylic acid group. This polymer would need a minimum NAVG MW of 5,000 daltons, giving an amine FGEW of 5,000 daltons [1 amine termination per 5,000 molecular weight (MW) of polymer], to be eligible for the exemption.

4.4.3   Exclusions for Elemental Criteria

A polymer manufactured under the 1995 rule must contain as an integral part of its composition at least two of the atomic elements of carbon, hydrogen, nitrogen, oxygen, sulfur, or silicon (C, H, N, O, S, Si). In addition to these six elements, only certain other elements are permitted as counterions or as an integral part of the polymer. The halogens (F, Cl, Br, and I) are allowed when covalently bonded to carbon. The monatomic counterions chloride, bromide, and iodide (Cl^-, Br^-, and I^-) are also permitted but not the fluoride anion (F^-). Other permitted monatomic ions are sodium, magnesium, aluminum, potassium, and calcium (Na^+, Mg^{2+}, Al^{3+}, K^+, and Ca^{2+}). The atomic elements lithium, boron, phosphorus, titanium, manganese, iron, nickel, copper, zinc, tin, and zirconium (Li, B, P, Ti, Mn, Fe, Ni, Cu, Zn, Sn, and Zr) are allowed at less than 0.20% by weight total, in any combination. No other elements are permitted, except as impurities.

4.4.4   Exclusions for Degradable or Unstable Polymers

A polymer is excluded from manufacture under the polymer exemption rule if the polymer is designed or reasonably expected to substantially degrade, decompose, or depolymerize. Polymers that could substantially decompose after manufacture or use, even if not intended to do so, are also ineligible. The following definition applies to this section.

> **Degradation, decomposition, or depolymerization**: chemical change in which a polymeric substance breaks down into simpler, smaller-weight substances by such actions as oxidation, hydrolysis, heat, sunlight, attack by solvents, or microbial action.

4.4.5   Exclusions by Reactants

Reactants and monomers may be present in a polymer at more than 2% by weight if they are either

(a)   on the TSCA Chemical Substance Inventory,

(b)   granted a §5 exemption (low-volume exemption, polymer exemption under the 1984 rule, etc.),

(c)   excluded from reporting, or

(d)   a nonisolated intermediate.

If the monomers and reactants do not fit one of these categories, then the polymer is ineligible for the polymer exemption. Both manufactured and imported polymers are covered by this rule. (See Section 4.7 for a discussion of the Two Percent Rule).

When a monomer or reactant is incorporated or charged at greater than 2%, it is considered part of the chemical identity of the polymer. Monomers and reactants that are not on the Inventory and do not have a §5 exemption may be used at less than 2% if they do not introduce any elements, properties, or functional groups that would render the polymer ineligible for exemption. The use of non-Inventory reactants or monomers applies effectively only to imported polymers; domestic manufacturers may not distribute or use any substance unless it is on the TSCA Inventory or exempt from TSCA reporting requirements. Non-Inventory monomers and reactants may be handled domestically only if they are intermediates generated in situ and are not isolated or if they are already exempt.

4.4.6 Exclusions for Water-Absorbing Polymers

Water-absorbing polymers with number-average molecular weight (NAVG MW) greater than or equal to 10,000 daltons are excluded from exemption. The preamble to the new rule describes how the EPA selected this NAVG MW and the level of water absorptivity for exclusion (2).

> **Water-absorbing polymer**: a polymeric substance that is capable of absorbing its weight in water.

4.4.7 Categories That Are No Longer Excluded from Exemption

Three exclusions have been dropped from the new polymer exemption rule. The three types of polymers that are no longer automatically excluded from the exemption are

(a) polymers containing less than 32% carbon,
(b) polymers manufactured from reactants containing halogen atoms (see Section 4.4.3) or cyano groups, and
(c) biopolymers.

These polymers must meet all of the criteria of the new rule in order to be manufactured under the exemption.

4.5   MEETING THE EXEMPTION CRITERIA

A new polymer that meets the definition of a polymer and is not automatically excluded must also meet one or more of the criteria listed in §723.250 (e)(1), (e)(2), or (e)(3) to be manufactured or imported under a polymer exemption.

4.5.1   The (e)(1) Exemption Criteria

For a new polymer to be manufactured under the (e)(1) exemption, it must meet the following requirements:

(a)   NAVG MW equal to or greater than 1,000 daltons and less than 10,000 daltons ($1,000 \leq$ NAVG MW $< 10,000$),

(b)   oligomer content less than 10% for molecular weight below 500 daltons, and

(c)   oligomer content less than 25% for molecular weight below 1,000 daltons.

Polymers must also comply with one of the following reactivity constraints in order to be eligible for exemption:

(a)   no reactive functional groups,

(b)   only low-concern functional groups, or

(c)   functional group equivalent weight (FGEW) above the threshold levels for moderate- and high-concern functional groups. (See Section 4.9 regarding calculation of the FGEW.)

Reactive functional group: an atom or group of atoms in a chemical substance that is intended or expected to undergo facile chemical reaction.

4.5.1.1   *Low-Concern Functional Groups and the (e)(1) Exemption*

Low-concern functional groups [see 40 CFR §723.250 (e)(1)(ii)(A)] may be used without limit. These functional groups are generally unreactive in biological settings. The low-concern functional groups are

- carboxylic acid groups,
- aliphatic hydroxyl groups,
- "ordinary" unconjugated olefinic groups (not specifically activated by being part of a larger functional group (i.e., vinyl ether) or by other activating influences (i.e., vinyl sulfone),
- butenedioic acid groups,
- conjugated olefinic groups contained in naturally occurring fats, oils, and carboxylic acids,
- blocked isocyanates, including ketoxime-blocked isocyanates,
- thiols,
- unconjugated nitrile groups, and
- halogens, with the exception of reactive halogen-containing groups such as benzylic or allylic halides.

Carboxylic esters, ethers, amides, urethanes, and sulfones are implicitly allowed because polyesters, polyethers, polyamides, polyurethanes, and polysulfones are permitted polymers under the exemption, as long as the reactivity of the functional groups has not been enhanced. For example, the dinitrophenyl ester of a carboxylic acid would not be allowed due to its enhanced reactivity.

In summary, a substance containing only low-concern functional groups may be manufactured under a polymer exemption if it

(a)   meets the definition of a polymer,
(b)   is not excluded,
(c)   has NAVG MW greater than or equal to 1,000 daltons and less than 10,000 daltons, and
(d)   meets the oligomer content criteria.

4.5.1.2 Moderate-Concern Functional Groups and the (e)(1) Exemption

Moderate-concern functional groups [see 40 CFR §723.250 (e)(1)(ii)(B)] may be used with functional group equivalent weight (FGEW) constraints. Each moderate-concern functional group present must have a FGEW greater than or equal to 1,000 daltons. If no high-concern groups are present, the combined FGEW ($FGEW_{combined}$) must be greater than or equal to 1,000 daltons. (See Section 4.8.2 to determine $FGEW_{combined}$). Moderate-concern reactive functional groups are

- acid halides,
- acid anhydrides,
- aldehydes,
- hemiacetals,
- methylolamides,
- methylolamines,
- methylolureas,
- alkoxysilanes bearing alkoxy groups greater than C2,
- allyl ethers,
- conjugated olefins, except those in naturally occurring fats, oils, and carboxylic acids,
- cyanates,
- epoxides,
- imines (ketimines and aldimines), and
- unsubstituted positions *ortho* and *para* to a phenolic hydroxyl group

In summary, a substance containing moderate-concern functional groups may be manufactured under a polymer exemption if it

(a) meets the definition of a polymer,
(b) is not excluded,
(c) has NAVG MW greater than or equal to 1,000 daltons and less than 10,000 daltons,
(d) meets the oligomer content criteria, and

(e)   has individual FGEWs and a $FGEW_{combined}$ greater than or equal to 1,000 daltons for moderate-concern groups when high-concern groups are not present. Low-concern groups may be present without limit.

4.5.1.3   High-Concern Functional Groups and the (e)(1) Exemption

Reactive groups not defined as low or moderate concern fall into the high-concern functional group category [see 40 CFR §723.250 (e)(1)(ii)(C)] and may be used with additional restrictions. If a polymer contains high-concern functional groups, each high-concern functional group present must meet a 5,000-dalton minimum permissible limit, each moderate-concern group present must meet the 1,000-dalton limit, and the polymer must have a $FGEW_{combined}$ of greater than or equal to 5,000 daltons. In calculating $FGEW_{combined}$, only moderate- and high-concern functional groups are considered. Low-concern groups are not included in the calculation and may be present without limit.

In summary, a substance containing high-concern functional groups may be manufactured under a polymer exemption if it

(a)   meets the definition of a polymer,
(b)   is not excluded,
(c)   has NAVG MW greater than or equal to 1,000 daltons and less than 10,000 daltons,
(d)   meets the oligomer content criteria,
(e)   has a $FGEW_{combined}$ greater than 5,000 daltons, and
(f)   meets the individual moderate- and high-concern FGEW limits of 1,000 and 5,000 daltons, respectively.

4.5.2   The (e)(2) Exemption Criteria

Polymers with NAVG MWs greater than or equal to 10,000 daltons are subject to §723.250 (e)(2). Oligomeric content must be less than 2% for species with molecular weight less than 500 daltons and less than 5% for species with molecular weight less than 1,000 daltons. No functional group restrictions apply to this exemption. The substance may not be excluded from exemption by any of the provisions listed previously [see 40 CFR §723.250(d)].

In summary, a substance may be manufactured under a polymer exemption if it

(a)   meets the definition of a polymer,

(b)   is not excluded,

(c)   has NAVG MW greater than or equal to 10,000 daltons, and

(d)   meets the oligomer content criteria.

4.5.3   The (e)(3) Exemption Criteria

Polyesters that have been prepared exclusively from a list of feedstocks identified in section (e)(3) of the new rule are eligible for exemption [see 40 CFR §723.250 (e)(3)]. Each monomer or reactant in the chemical identity of the polymer (charged at any level) must be on the list in order for the polymer to be eligible for exemption. Any substance on the list that is not on the TSCA Inventory may not be used in domestic manufacture.

Polyesters that are exempt under (e)(3) must meet the definition of a polymer and must not be excluded from exemption by other factors. Biodegradable polyesters, for example, would be excluded from an (e)(3) exemption.

Table 4.3 lists the monomers and reactants that may be used in preparing a polyester that is eligible for an (e)(3) exemption.

4.6   NUMERICAL CONSIDERATIONS

Several numerical criteria must be considered when deciding if a polymer is eligible for an exemption:

(1)   Number-average molecular weight (NAVG MW) is used to determine whether an eligible polymer meets an (e)(1) or (e)(2) exemption or whether a water-absorbing polymer is excluded from exemption by the 10,000-dalton restriction.

(2)   The "Two Percent Rule" prescribes whether a monomer or other reactant is part of the chemical identity of a polymer. If

TABLE 4.3. The (e)(3) Monomer and Reactant List (in order by CAS Registry Number)

[56-81-5]	1,2,3-Propanetriol
[57-55-6]	1,2-Propanediol
[65-85-0]	Benzoic acid
[71-36-3]†	1-Butanol
[77-85-0]	1,3-Propanediol, 2-(hydroxymethyl)-2-methyl-
[77-99-6]	1,3-Propanediol, 2-ethyl-2-(hydroxymethyl)-
[80-04-6]	Cyclohexanol,4,4'-(1-methylethylidene)bis-
[88-99-3]	1,2-Benzenedicarboxylic acid
[100-21-0]	1,4-Benzenedicarboxylic acid
[105-08-8]	1,4-Cyclohexanedimethanol
[106-65-0]	Butanedioic acid, dimethyl ester
[106-79-6]	Decanedioic acid, dimethyl ester
[107-21-1]	1,2-Ethanediol
[107-88-0]	1,3-Butanediol
[108-93-0]	Cyclohexanol
[110-15-6]	Butanedioic acid
[110-17-8]	2-Butenedioic acid (E)-
[110-40-7]	Decanedioic acid, diethyl ester
[110-63-4]	1,4-Butanediol
[110-94-1]	Pentanedioic acid
[110-99-6]	Acetic acid, 2,2'-oxybis-
[111-14-8]	Heptanoic acid
[111-16-0]	Heptanedioic acid
[111-20-6]	Decanedioic acid
[111-27-3]	1-Hexanol
[111-46-6]	Ethanol, 2,2'-oxybis-
[112-05-0]	Nonanoic acid
[112-34-5]	Ethanol, 2-(2-butoxyethoxy)-
[115-77-5]	1,3-Propanediol, 2,2-bis(hydroxymethyl)-
[120-61-6]	1,4-Benzenedicarboxylic acid, dimethyl ester
[121-91-5]	1,3-Benzenedicarboxylic acid
[123-25-1]	Butanedioic acid, diethyl ester
[123-99-9]	Nonanedioic acid
[124-04-9]	Hexanedioic acid
[126-30-7]	1,3-Propanediol, 2,2-dimethyl-
[141-28-6]	Hexanedioic acid, diethyl ester
[142-62-1]	Hexanoic acid
[143-07-7]	Dodecanoic acid
[144-19-4]	1,3-Pentanediol, 2,2,4-trimethyl-

TABLE 4.3. *(Continued)*

[505-48-6]	Octanedioic acid
[528-44-9]	1,2,4-Benzenetricarboxylic acid
[624-17-9]	Nonanedioic acid, diethyl ester
[627-93-0]	Hexanedioic acid, dimethyl ester
[629-11-8]	1,6-Hexanediol
[636-09-9]	1,4-Benzenedicarboxylic acid, diethyl ester
[693-23-2]	Dodecanedioic acid
[818-38-2]	Pentanedioic acid, diethyl ester
[1119-40-0]	Pentanedioic acid, dimethyl ester
[1459-93-4]	1,3-Benzenedicarboxylic acid, dimethyl ester
[1732-08-7]	Heptanedioic acid, dimethyl ester
[1732-09-8]	Octanedioic acid, dimethyl ester
[1732-10-1]	Nonanedioic acid, dimethyl ester
[1852-04-6]	Undecanedioic acid
[2163-42-0]	1,3-Propanediol, 2-methyl
[3302-10-1]	Hexanoic acid, 3,3,5-trimethyl-
[8001-20-5]*	Tung oil
[8001-21-6]*	Sunflower oil
[8001-22-7]*	Soybean oil
[8001-23-8]*	Safflower oil
[8001-26-1]*	Linseed oil
[8001-29-4]*	Cottonseed oil
[8001-30-7]*	Corn oil
[8001-31-8]*	Coconut oil
[8002-50-4]*	Fats and glyceridic oils, menhaden
[8016-35-1]*	Fats and glyceridic oils, oiticica
[8023-79-8]*	Palm kernel oil
[8024-09-7]*	Oils, walnut
[13393-93-6]	1-Phenanthrenemethanol, tetradecahydro-1,4a-dimethyl-7-(1-methylethyl)-
[25036-25-3]	Phenol,4,4′-(1-methylethylidene)bis-, polymer with 2,2′-[(1methylethylidene)bis(4,1-phenyleneoxymethylene)]-bis[oxirane]
[25119-62-4]	2-Propen-1-ol, polymer with ethenylbenzene
[25618-55-7]	1,2,3-Propanetriol, homopolymer
[61788-47-4]*	Fatty acids, coco
[61788-66-7]*	Fatty acids, vegetable oil
[61788-89-4]*	Fatty acids, C18-unsaturated, dimers
[61789-44-4]*	Fatty acids, castor oil
[61789-45-5]*	Fatty acids, dehydrated castor oil

TABLE 4.3. (*Continued*)

[61790-12-3]*	Fatty acids, tall oil
[67701-08-0]*	Fatty acids, C16-18 and C18-unsaturated
[67701-30-8]*	Glycerides, C16-18 and C18-unsaturated
[68037-90-1]*	Silsesquioxanes, PhPr
[68132-21-8]*	Oils, perilla
[68153-06-0]*	Fats and glyceridic oils, herring
[68308-53-2]*	Fatty acids, soya
[68424-45-3]*	Fatty acids, linseed oil
[68440-65-3]*	Siloxanes and silicones, di-Me, di-Ph, polymers with Ph silsesquioxanes, methoxy-terminated
[68957-04-0]*	Siloxanes and silicones, di-Me, methoxy Ph, polymers with Ph silsesquioxanes, methoxy-terminated
[68957-06-2]*	Siloxanes and silicones, MePh, methoxy Ph, polymers with Ph silsesquioxanes, methoxy- and Ph-terminated
[72318-84-4]*	Methanol, hydrolysis products with trichlorohexylsilane and trichlorophenylsilane
[84625-38-7]*	Fatty acids, sunflower oil
[68649-95-6]*	Linseed oil, oxidized
[68953-27-5]*	Fatty acids, sunflower oil, conjugated
[91078-92-1]*‡	Fats and glyceridic oils, babassu
[93165-34-5]*‡	Fatty acids, safflower oil
[93334-41-9]*‡	Fats and glyceridic oils, sardine
[120962-03-0]*	Canola oil
[128952-11-4]*‡	Fats and glyceridic oils, anchovy

[No Registry #]*‡ Fatty acids, tall oil, conjugated.
[No Registry #]*‡ Oils, cannabis.
*Designates chemical substances of unknown or variable composition, complex reaction products, or biological materials (UVCB substances). The CAS Registry Numbers for UVCB substances are not used in Chemical Abstracts and its indexes.
†1-Butanol may not be used in a substance manufactured from fumaric or maleic acid because of potential risks associated with esters which may be formed by reaction of these reactants.
‡Not on the TSCA Inventory.

a monomer or reactant is charged or incorporated at greater than 2% composition, it must be on the TSCA Inventory, excluded from reporting, or otherwise exempt under section 5 of TSCA. For (e)(1) and (e)(2) exemptions, imported polymers may have non-Inventory monomers or reactants present at less than or equal to 2%. Domestic manufacture under an (e)(1) or (e)(2) exemption requires all monomers and reactants to either be

(a) on the Inventory,
(b) a nonisolated intermediate,
(c) otherwise exempt, or
(d) excluded from reporting.

For (e)(3) polymers, all monomers and reactants must be from the (e)(3) list, *regardless* of charge. Only those charged at greater than 2% will be part of the identity of the polymer. The Two Percent Rule is explained in greater detail in Section 4.7.

(3) The functional group equivalent weight (FGEW) is a measure of the concentration of moderate- and high-concern functional groups in a polymer. FGEW is used to determine the exemption eligibility of a polymer with NAVG MW between 1,000 and 10,000 daltons, as well as the eligibility of cationic polymers for exclusion. Determination of FGEW is detailed in Section 4.9.

4.6.1 Calculating Number-Average Molecular Weight

Two methods are commonly used to calculate the molecular weight of a polymer: number-average molecular weight (NAVG MW) and weight-average molecular weight (WAVG MW). The Agency uses NAVG MW to determine eligibility for the polymer exemption rule. NAVG MW represents the average weight of the predominant components of a polymer sample by taking into account the number of molecules of different molecular weights present in the sample. With WAVG MW, a few large molecules can bias the average and present an inaccurate picture of the majority of molecules in the sample.

Number-average molecular weight (M_n) and weight-average molecular weight (M_w) may be calculated using eqs. 4.1 and 4.2, respectively (5). In the equations, N_i is the number of molecules at a given molecular weight and M_i is the molecular weight of a given polymer fraction.

$$M_n = \frac{\sum_{i=1}^{\infty} N_i M_i}{\sum_{i=1}^{\infty} N_i} \qquad \text{(Eq. 4.1)}$$

$$M_w = \frac{\sum_{i=1}^{\infty} N_i M_i^2}{\sum_{i=1}^{\infty} N_i M_i} \qquad \text{(Eq. 4.2)}$$

Example 11

The numerical difference obtained when using NAVG MW versus WAVG MW is best illustrated with an example. Suppose a polymer contains 200 molecules that weigh 1000 daltons, 300 molecules that weigh 1500 daltons, 400 molecules that weigh 2000 daltons, and 2 molecules that weigh 1,000,000 daltons. For this polymer, NAVG MW = 3825 daltons and WAVG MW = 5.80×10^5 daltons.

$$M_n = \frac{(200)(1000) + (300)(1500) + (400)(2000) + (2)(1,000,000)}{200 + 300 + 400 + 2}$$

$$= 3825$$

$$M_w = \frac{(200)(1000)^2 + (300)(1500)^2 + (400)(2000)^2 + (2)(1,000,000)^2}{(200)(1000) + (300)(1500) + (400)(2000) + (2)(1,000,000)}$$

$$= 5.80 \times 10^5$$

Of these two calculations, NAVG MW more accurately represents the majority of the molecules in the polymer sample. The WAVG MW is significantly higher due to the presence of the two molecules weighing 1,000,000 daltons. The Agency requires the manufacturer of an exempt polymer to keep records of the "lowest" NAVG MW at which the polymer is to be made. This value is obtained by analyzing samples for a series of batches in the production of the polymer. The lowest NAVG MW value is needed to determine eligibility under the (e)(1) and (e)(2) criteria.

4.6.2   Analytical Techniques for Determination of NAVG MW

A number of analytical techniques may be used to determine NAVG MW (3, 6, 7). A brief summary of these techniques is provided below, although any verifiable method is acceptable for purposes of polymer exemption. The most commonly used methods, and the basis for their utility, are

- gel permeation chromatography (polymer size),
- membrane osmometry (colligative property),
- vapor-phase osmometry (colligative property),
- vapor pressure lowering (colligative property),
- ebulliometry (colligative property),
- cryoscopy (colligative property), and
- end-group analysis (chemical reactivity).

4.6.2.1   Gel Permeation Chromatography

Gel permeation chromatography (GPC) is the most frequently used and generally most reliable method for determining NAVG MW of polymers. The polymer sample separates according to size as it traverses a column packed with a porous material, such as polyacrylamide. Smaller molecules enter the pores of the packing material, eluting after the larger molecules. The column is first calibrated using polymers of known weight and similar structure. Refractive index and ultraviolet absorption are used to detect the polymers as they elute from the column.

Band broadening is problematic when measuring low molecular weight polymers or when columns are unevenly packed or contain dead volumes. Empirical calibrations of the instrument can be made to minimize broadening (8) but become unimportant when the ratio of the WAVG MW to the NAVG MW is greater than two. Many high molecular weight polymers are insoluble in suitable GPC solvents, limiting the application of this technique to certain polymers.

4.6.2.2   Membrane Osmometry

The principle of osmosis is used to determine NAVG MW by membrane osmometry (9). A semipermeable membrane separates the solvent from the polymer. The solvent passes through the membrane until equilibrium is reached. The pressure differential that is created depends on both the concentration difference and the molecular weight of the polymer. One problem associated with this technique is the diffusion of low-weight oligomers through the membrane, affecting both the accuracy and reliability of this method. Diffusion is generally not a problem, however, for unfractionated polymers with NAVG MWs greater than 50,000 daltons. OECD guidelines (3) cite an upper limit NAVG MW of 200,000 daltons that may be measured with confidence when using membrane osmometry.

4.6.2.3   Vapor-Phase Osmometry

Vapor-phase osmometry is most accurate for polymers with NAVG MW less than 20,000 daltons (OECD guidelines) (3). This method compares the evaporation rates for a solvent aerosol with at least three other aerosols of varying polymer concentrations in the same solvent. The evaporation rate differs between the pure solvent and the polymer solution, giving rise to a small temperature difference, which is proportional to the NAVG MW. This technique is applicable to polymer samples whose molecular weight is too low to be measured by membrane osmometry.

4.6.2.4   Vapor Pressure Lowering

Vapor pressure lowering measures the difference in vapor pressure between the pure solvent and a solution of the polymer in the solvent. The vapor pressure of the pure solvent is compared with the vapor pressure of at least three concentrations of the polymer mixed with the same solvent. This method is of limited utility for polymers up to 20,000 NAVG MW.

4.6.2.5   Ebulliometry

Ebulliometry employs the boiling point elevation of a polymer solution to determine NAVG MW (10). Although this technique is accurate for polymers up to 30,000 NAVG MW, results may be affected by the foaming that occurs when some polymers are boiled. The larger surface area of the foam causes the polymer to concentrate in the foam, decreasing the measured concentration of the polymer in solution. The ebulliometer may be calibrated with a substance of known molecular weight, such as octacosane, which has a molecular weight of 396 daltons.

4.6.2.6   Cryoscopy

Cryoscopy utilizes freezing point depression to determine NAVG MW. Controlled crystallization of the solvent may be facilitated by use of a nucleating agent, although care must be taken to avoid supercooling. A substance of known molecular weight is used for calibration. The technique is accurate for molecular weight up to 30,000 daltons.

4.6.2.7   End-Group Analysis

The end-group analysis method requires basic information about the polymer, such as the overall structure and the chain-terminating end groups. The NAVG MW is determined by considering the number of molecules in a given weight of a sample. This technique works best with linear condensation polymers but is less useful for branched condensation polymers and addition polymers. End-group analysis may be applied to branched polymers when the polymerization kinetics are well known, allowing the degree of branching to be estimated from the amount of feedstock charged. The situation for addition polymers is more complicated. Molecular weight can be determined through end-group analysis by analyzing for specific initiator fragments containing identifiable functional groups, elements, or radioactive atoms; for chain-terminating groups arising from transfer reactions with solvent; or for unsaturated end groups, such as in polyethylene or poly-α-olefins.

Several analytical techniques are used to distinguish end groups from the main polymer skeleton. Nuclear magnetic resonance, titration, and derivatization are most commonly used. The carboxyl end groups in polyesters, for example, may be titrated with a base. When the polymer cannot be titrated due to insolubility, however, infrared spectroscopy may prove useful. End-group analysis is a valid technique for NAVG MWs up to 50,000 daltons.

4.7 THE TWO PERCENT RULE AND CHEMICAL IDENTITY

A polymer is not eligible for exemption [see 40 CFR §723.250(d)(4)] if it contains monomers and/or reactants at greater than 2% by weight that are not

(1) included on the TSCA Inventory,
(2) manufactured under an applicable TSCA §5 exemption,
(3) excluded from exemption, or
(4) a nonisolated intermediate.

Monomers and reactants at greater than 2% by weight constitute the "chemical identity" of the polymer. Monomers and reactants at less than or equal to 2% by weight of an exempt polymer are not considered part of the "chemical identity"; the use of such monomers and reactants is discussed below.

A manufacturer or importer of an exempt polymer is restricted by the weight percent of each monomer or reactant in two major ways.

(1) If a certain monomer or reactant is used at less than or equal to 2% by weight in an exempt polymer, it may not later be used at greater than 2% (under the exemption for the same polymer). The chemical identity of the polymer would change if greater than 2% of the monomer or reactant were used.

(2) If a monomer or reactant is used at greater than 2% by weight in an exempt polymer, it may not be eliminated completely from the polymer (under the exemption for the same polymer). Again, this would change the chemical identity of the polymer.

If the manufacturer makes one of the identity-changing modifications mentioned above, the manufacturer must do one of the following:

(1) find the new polymer identity on the TSCA Inventory,

(2) submit a premanufacture notification (PMN) at least 90 days before manufacture if the new polymer is not on the Inventory, or

(3) meet the conditions of a PMN exemption to cover the new polymer identity.

Non-Inventory monomers and reactants cannot be used at any level in domestic manufacture, unless subject to another §5 exemption or used as nonisolated intermediates. Inventory-listed reactants and monomers at less than or equal to 2% may be exchanged under one exemption, as long as the polymer remains eligible for exemption and the manufacturer maintains records of such changes, as stipulated in the rule. An exempt imported polymer may contain non-Inventory monomers and reactants at less than or equal to 2% by weight, as long as these substances do not otherwise render the polymer ineligible for exemption. Recall, however, the (e)(3) exemption, which stipulates that monomers and reactants must only come from the (e)(3) list, regardless of level used. For all polymer types, the "Two Percent Rule" must be applied in conjunction with other restrictions.

> **Percent by weight**: the weight of the monomer or other reactant used as a percentage of the dry weight of the polymer.

The Agency believes that the actual content of a polymer provides the best indication of its physical, chemical, and toxicological properties. EPA recognizes, however, that it is more convenient to use the amount of reactant charged to the reactor, rather than the amount of reactant incorporated into the polymer, when calculating the percentage of each reactant present. Under the 1995 PMN rule revisions, the Agency now accepts two methods for determining the "percent by weight" of each reactant for the purpose of establishing the chemical identity of a polymer.

(1)   The percent charged method: the percent composition of each monomer or reactant is established by the amounts charged to the reaction vessel.

(2)   The percent incorporated method: the percent composition is based on the minimum theoretical amount of monomer or reactant needed to be charged to the reactor to account for the amount analytically determined to be incorporated into the polymer. The percent composition of each monomer or reactant whose fragment is present in the polymer may be established by theoretical calculations if it can be documented that an analytical determination cannot or need not be made to comply with the new polymer exemption rule.

The Agency specifies the polymer identity information that must be kept by the manufacturer or importer [see 40 CFR §723.250(g)]. A manufacturer or importer must identify the specific chemical identity and CAS registry number (or EPA accession number) for each monomer or reactant used at *any* weight in the manufacture of an exempt polymer. A CAS registry number may not be necessary in cases in which a monomer or reactant has been exempted or excluded from exemption as the subject of a previous PMN. Manufacturers and importers are advised to maintain known CAS registry numbers on file for all monomers and reactants.

If possible, a structural diagram of the polymer should be provided, as the structure best represents the intended identity of the substance. Consider the reaction between two feedstocks, one a carboxylic acid and the other an amine. The feedstocks can combine to form an amide or carboxylic acid salts. A structure clearly indicates the product intended by the manufacturer or importer. Note that all monomers and reactants at greater than 2% by weight in the polymer should be included in the polymer structural diagram.

4.7.1   Percent Charged Method

The percent by weight charged is determined by dividing the grams of feedstock charged (GFC) to the reactor by the grams of dry polymer formed (GPF). GPF is the dry weight of the polymer isolated from the reaction.

$$\text{Percent by weight charged} = \frac{\text{(GFC)}}{\text{(GPF)}} \times 100 \qquad \text{(Eq. 4.3)}$$

The sum of the weights of the reactants charged may significantly exceed 100% if the monomers or reactants lose a substantial portion of their structure when incorporated into the polymer. The polymerization of vinyl acetate to polyvinyl alcohol (Fig. 4.6) illustrates the problem encountered in this type of calculation.

Example 12

Vinyl acetate has a molecular weight of 86 daltons, whereas the repeating unit in the polymer —CH_2—$CH(OH)$— has a molecular weight of 44 daltons. In this case, the GFC-to-GPF ratio is the same as the feedstock MW-to-fragment MW ratio because a single monomer is the sole source of the repeating unit in the polymer. The equation may be simplified to

$$\text{Wt.\% of vinyl acetate} = \frac{(86)}{(44)} \times 100 = 195$$

With the use of the percent charged method, the weight percent vinyl acetate charged to the reactor is 195%.

4.7.2 Percent Incorporated Method

The 1995 PMN rule amendments state that the weight percent is based on "the minimum weight of monomer or other reactant required in theory to account for the actual weight of monomer or other reactant molecule or fragments chemically incorporated (chemically com-

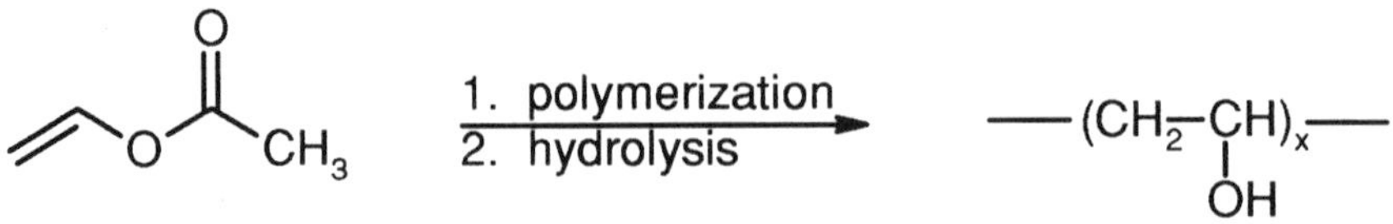

Figure 4.6 Polymerization of vinyl acetate to polyvinyl alcohol (Example 12).

bined) in the polymeric substance manufactured (2)." In other words, weight percent is determined by calculating the theoretical weight of monomer or reactant necessary to produce the actual weight of monomer or reactant incorporated into the polymer. The degree of incorporation of the fragment resulting from the monomer or reactant must be measured to calculate the percent incorporated.

In some cases, it may not be possible to determine analytically the degree of incorporation for every type of reactant. This is true for random polymerizations, in which no repeating subunits exist, and for polymerizations, in which the structures of the reactants are not completely specified, such as conjugated sunflower oil fatty acids. A structural unit within the polymer that corresponds to the specific monomer from which it originated must be identified. Different monomers can act as the source of the same monomer unit. Complete or efficient incorporation of reactants into the polymer cannot be assumed. If the percent incorporated cannot reliably be measured or estimated, the manufacturer must use the percent charged method.

The following information must be known to calculate the weight percent incorporated for a reactant:

- molecular weight of reactant charged,
- molecular weight of the fragment incorporated into the polymer, and
- analytically determined amount of reactant incorporated into the polymer.

These data permit determination of the number of moles of fragment present in the polymer (eq. 4.4), which is proportional to the amount of feedstock that reacted to form the polymer.

$$\frac{\text{Wt\% frag}}{\text{MW frag}} = \frac{(\text{g frag})}{(100 \text{ g polymer})} \times \frac{(\text{mol frag})}{(\text{g frag})}$$

$$= \frac{\text{mol frag}}{100 \text{ g polymer}} = \text{Ratio A} \qquad \text{(Eq. 4.4)}$$

The weight percent of reactant incorporated is calculated using eq. 4.5. *Ratio A* is first converted to moles of reactant per 100 g of polymer and then multiplied by the molecular weight of the reactant to yield the weight percent of reactant incorporated.

$$\text{Wt\% reactant incorp.} = (\text{Ratio A}) \times \frac{(\text{mol reactant})}{(\text{mol frag})}$$
$$\times (\text{MW reactant}) \times 100 \qquad (\text{Eq. 4.5})$$

Example 13

The polymerization of ethylene glycol with dimethyl terephthalate may be used to illustrate the application of the percent incorporated method. The polyester formed incorporates both oxygen atoms from the ethylene glycol, with concurrent loss of two methoxy groups from the terephthalate ester. Empirically, the polymer was shown to contain 13.2% by weight of the terephthaloyl unit ($C_8H_4O_2$), which has a molecular weight of 132 daltons. *Ratio A* for the terephthaloyl unit may be calculated using eq. 4.4.

$$\text{Ratio A} = \frac{(13.2 \text{ g})}{(100 \text{ g polymer})} \times \frac{(1 \text{ mol})}{(132 \text{ g})} = \frac{(0.10 \text{ mol terephthaloyl})}{(100 \text{ g polymer})}$$

Each mole of dimethyl terephthalate (MW = 194 daltons) produces one mole of fragment, giving a molar conversion factor of 1. Inserting this data into eq. 4.5 gives a weight percent of reactant incorporated of 19.4. This value means that dimethyl terephthalate must be charged to the reaction vessel at 19.4% in order for the terephthaloyl group to be incorporated at 13.2%.

$$\text{Wt\% reactant incorp.} = \frac{(0.10 \text{ mol})}{(100 \text{ g polymer})} \times \frac{(1 \text{ mol feedstock})}{(1 \text{ mol fragment})}$$
$$\times 194 \times 100 = 19.4$$

Example 14

Consider the reaction above in which diethyl terephthalate (MW = 222 daltons) is used in place of dimethyl terephthalate. The calculations are identical with the exception of the molecular weight of the reactant, which increases from 194 to 222 daltons. The weight percent of reactant incorporated climbs to 22.2% with diethyl terephthalate. More diethyl terephthalate needs to be charged to the reaction vessel than dimethyl terephthalate to achieve the same degree of incorporation of the terephthaloyl unit. These values make sense because of the size of the alkoxy groups that are lost during the polymerization reaction; the ethoxy group is larger than the methoxy group.

Example 15

Neutralizers are often used in excess over the amount actually incorporated into a polymer. If the amount incorporated is 2% or less, the neutralizer may be omitted from the identity of the polymer. For example, a polymer containing free carboxylic acid functional groups was neutralized with sodium hydroxide (NaOH, formula weight = 40 g/mol). If NaOH is charged to the reactor at 10%, is it part of the identity of the polymer? Analysis of the polymer salt indicated that the polymer contained 0.92% by weight of sodium (atomic weight = 23 g/mol), coming only from the base. This amount of sodium corresponds to 0.04 mol of sodium per 100 g of polymer, or 1.6 g of NaOH per 100 g of polymer. In other words, 1.6% by weight of NaOH was incorporated into the polymer, although 10% NaOH was charged to the reactor vessel. The polymer does not need to be described as the sodium salt because the weight percent of NaOH incorporated is less than 2%.

$$\text{Ratio A for Na} = \frac{(0.92 \text{ g Na})}{(100 \text{ g polymer})} \times \frac{(1 \text{ mol Na})}{(23 \text{ g Na})} = \frac{(0.04 \text{ mol Na})}{(100 \text{ g polymer})}$$

$$\text{Wt\% incorp. NaOH} = \frac{(0.04 \text{ mol Na})}{(100 \text{ g polymer})} \times \frac{(40 \text{ g NaOH})}{(1 \text{ mol NaOH})} \times 100 = 1.6$$

In the example above, suppose that sodium bicarbonate (NaHCO$_3$, formula weight = 84 g/mol) was used as the neutralizing agent in place of sodium hydroxide. The same number of moles of sodium per 100 g of polymer translates to 3.36% by weight sodium bicarbonate due to the higher formula weight of the base. In this case, the polymer must be described as the sodium salt because the weight percent of $NaHCO_3$ is greater than 2%.

$$\text{Wt\% incorp. } NaHCO_3 = \frac{(0.04 \text{ mol Na})}{(100 \text{ g polymer})} \times \frac{(84 \text{ g } NaHCO_3)}{(1 \text{ mol } NaHCO_3)}$$

$$\times \ 100 = 3.36$$

If a combination of bases is used for neutralization, the amounts incorporated should be adjusted according to the mole ratios of the neutralizing agents charged, provided that the reactivities of the bases are similar. When the reactivities of the bases are not equal, assume that the most reactive neutralizing agent is consumed first, the second most reactive consumed next, and so on.

Example 16

Calculating the weight percent incorporated of an initiator is similar to that for excess neutralizing base, as the amount of initiator charged may be larger than the amount of initiator incorporated into the polymer. The initiator will not be included in the chemical identity of the polymer if it is incorporated at 2% or less. Polymers with PMNs and Notice of Commencement (NOCs) may include or omit the initiator at the discretion of the submitter. If an initiator is not in the identity of an exempted polymer or in the identity of a polymer covered by a PMN or NOC, then the initiator may be changed without establishing another polymer exemption or submitting another PMN, provided that the alternate initiator remains at less than 2% and does not exclude the polymer in other ways.

Consider, for example, a polyolefin with NAVG MW of 9000 daltons, synthesized using azobisisobutyronitrile (AIBN; MW = 164 g/mol), charged at 3%. Each mole of AIBN dissociates to produce 2 mol of nitrile radicals (CN, formula weight = 26 g/mol). The polymer sample was found to contain 0.29% nitrile by weight; the nitrile moiety was assumed to originate from AIBN. Converting grams to moles yields 0.011 mol of fragment per 100 g of polymer (0.29 g $\div$ 26 g/mol = 0.011 mol). Because 1 mol of reactant produces 2 mol of product, the weight percent of reactant incorporated may be calculated using eq. 4.5:

$$\text{Wt\% AIBN incorp.} = \frac{(0.011 \text{ mol frag})}{(100 \text{ g polymer})} \times \frac{(1 \text{ mol reactant})}{(2 \text{ mol frag})}$$
$$\times (164)(100) = 0.9$$

Although AIBN was charged to the reaction vessel at 3% by weight, only 0.9% by weight was incorporated into the polymer. Therefore, the submitter does not need to include AIBN in the polymer identity, as the weight percent is less than 2%.

4.8 METHODS FOR DETECTION OF POLYMER COMPOSITION

Many methods are available for the chemical analysis of polymers and the determination of the weight percent of fragments incorporated. Any verifiable method of analysis is acceptable. The more common methodologies used include

- classical chemical analysis (elemental analysis, titration, etc.),
- mass spectrometry,
- gas chromatography,
- infrared spectroscopy,
- nuclear magnetic resonance, and
- X-ray diffraction analysis

4.8.1 Mass Spectrometry

In mass spectrometry, a sample is bombarded with high-energy electrons, creating radical cations that fragment into smaller pieces. The fragments are separated according to their mass-to-charge ratio in a magnetic field. The structure and composition of the parent compound can be deduced from the abundance of the ionic species formed. Commonly, polymers are thermally degraded into low molecular weight fragments before being injected into the mass spectrometer.

4.8.2   Gas Chromatography

In gas chromatography (GC), the vaporized sample partitions between a mobile gas phase and a stationary liquid phase or solid adsorbent. Separation is achieved as each component spends a different amount of time adsorbed to the stationary phase. The peaks in the gas chromatogram are proportional to the concentration of the components, providing information on the number, type, and weight percentage of each species present.

4.8.3   Infrared Spectroscopy

Sample absorption of infrared radiation gives rise to molecular vibration spectra. The infrared spectra of polymers can be surprisingly simple because many of the normal vibrations have the same frequency and the strict selection rules for absorption prevent many of the vibrations from causing absorption peaks in the spectrum. However, infrared spectroscopy is rarely used for quantitative analysis of polymers.

4.8.4   Nuclear Magnetic Resonance Spectroscopy

Nuclear magnetic resonance (NMR) is a powerful tool for studying chain configuration, sequence distribution, and microstructures in polymers. NMR utilizes the property of spin angular momentum possessed by nuclei with odd mass numbers or atomic numbers. The sample is placed in a strong magnetic field, causing nuclei to align with or against the applied field. Irradiation of the sample with electromagnetic energy causes the low-energy nuclei to spin-flip to the high-energy state. At this point, the nuclei are said to be "in resonance" with the applied radiation. Each unique nucleus absorbs the electromagnetic energy at a different field strength, resulting in a series of peaks; the position, multiplicity, and intensity of the peaks provide information about the structure of the polymer.

4.8.5   X-Ray Diffraction Analysis

X-ray diffraction may be used to detect the presence of structures that are arranged in an orderly array or lattice. The interaction between the lattice and electromagnetic radiation provides information on the geometry of the structures. A fiber drawn from the polymer is typically used to determine crystal structure, as single crystals of polymer are generally too small for X-ray diffraction.

4.9 CALCULATING FUNCTIONAL GROUP EQUIVALENT WEIGHT

When reactive functional groups are present in a polymer at greater than or equal to 2%, the minimum functional group equivalent weight (FGEW) requirements for the exemption category under which the polymer is manufactured or imported must be met. Polymers that are exempt under the (e)(1) criteria must meet or exceed the minimum permissible equivalent weights for reactive functional groups. There are no functional group restrictions for polymers meeting the (e)(2) exemption except for cationic and potentially cationic groups. For (e)(3) polymers, reactive functional groups of moderate and high concern would not be present in any polymer derived from the monomers on the allowed list. Monomers on the list are not expected to contain reactive functional groups of moderate or high concern once they are incorporated into the polymer. Therefore, the (e)(3) section of the new polymer exemption rule does not have any FGEW requirements.

Table 4.4 lists the allowable thresholds for certain reactive functional groups for (e)(1) polymers. (Note: This is not an exhaustive list. Consult the 1995 polymer exemption rule for groups not mentioned.) If a functional group is not mentioned in the rule among the low- or moderate-concern groups, it is considered to be a high-concern functional group. Low-concern reactive groups may be used without limit, and no thresholds have been set for them.

If a FGEW cannot be determined empirically by recognized, scientific methodology, a worst-case estimate must be made for the FGEW, in which all moderate- and high-concern functional moieties must be factored. Limited guidance on calculating FGEWs is provided below. The following sections cover methods based on end-group analysis, percent charged, and nomograph use.

TABLE 4.4. Allowable Thresholds for Selected Reactive Functional Groups

---

Moderate Concern: The minimum permissible FGEW is 1000 daltons.
Acid halides
Acid anhydrides
Aldehydes
Alkoxysilanes where alkyl is greater than C2
Allyl ethers
Conjugated olefins
Cyanates
Epoxides
Hemiacetals
Hydroxymethylamides
Imines
Methylolamides
Methylolamines
Methylolureas
Unsubstituted position *ortho* or *para* to phenolic hydroxyl

High Concern: The minimum permissible FGEW is 5000 daltons. *
Acrylates
Alkoxysilanes, where alkyl = methyl or ethyl
Amines
Aziridines
Carbodiimides
Halosilanes
Hydrazines
Isocyanates
Isothiocyanates
α-Lactones; β-lactones
Methacrylates
Vinyl sulfones

---

*For polymers containing high-concern functional groups, the combined functional group equivalent weight ($FGEW_{combined}$) must be greater than or equal to 5000 daltons taking into account high-concern (e)(1)(ii)(c) and, if present, moderate-concern (e)(1)(ii)(b) functional groups.

4.9.1 End-Group Analysis

Reactive functional groups are typically found only at the chain ends of condensation polymers, such as polyesters and polyamides. All other reactive functional groups in the monomers combine to form the polymer backbone, leaving only the termini of the chain with reactive functional groups. For this type of polymer, the FGEW may be determined by theoretical end-group analysis, regardless of the nature of the reactive functional group.

Linear polymers are synthesized from monomers containing two reactive functional groups. The FGEW for a linear polymer, with either nucleophilic or electrophilic reagents in excess, is one-half the NAVG MW.

Example 17

A polyamide, NAVG MW = 1000 daltons, is synthesized from excess ethylene diamine (two nucleophiles) and adipic acid (two electrophiles). The polymer is expected to be amine-terminated at both ends due to the use of excess amine. The amine equivalent weight would be one-half the NAVG MW or 500 daltons.

Determination of the FGEW for branched polymers is more complex. For a simple, branched condensation polymer, synthesized from one monomer with more than two reactive sites, the FGEW must be calculated from the total number of end groups present in the polymer. This number is determined by estimating the degree of branching, which is derived from the number of reactive functional groups in the monomer. It is assumed that the monomer responsible for branching will be consumed in its entirety to form the polymer.

The FGEW for simple, branched polymers may be estimated using eqs. 4.6–4.9. The equivalent weight of the monomer is the molecular weight of the monomer divided by the weight percent charged to the reaction vessel (eq. 4.6). A monomer equivalent weight of 1000 daltons means that there is 1 mol of monomer for every 1000 daltons of polymer.

$$\text{Monomer equivalent weight} = \frac{\text{(Monomer molecular weight)}}{\text{(Weight percent charged)}}$$

$$(\text{Eq. 4.6})$$

Monomer equivalent weight is used to determine the degree of branching (eq. 4.7). First, the NAVG MW is divided by the monomer equivalent weight. The degree of branching is obtained by multiplying this result by the number of reactive groups that are *not* used to make the polymer backbone (NRG − 2). NRG corresponds to the number of reactive groups originally in the monomer.

$$\text{Degree of branching} = \frac{\text{(NAVG MW)}}{\text{(Monomer equivalent weight)}} \times (\text{NRG} - 2)$$

(Eq. 4.7)

The total number of end groups in the polymer is simply the degree of branching plus two, where two represents the number of end groups of the polymer backbone. Equation 4.8 is the following: total number of polymer end groups = degree of branching + 2.

Finally, the FGEW is calculated by dividing the NAVG MW by the number of end groups in the polymer.

$$\text{FGEW} = \frac{\text{NAVG MW}}{\text{Total number of polymer end groups}}$$

(Eq. 4.9)

Example 18

A polymer is manufactured from pentaerythritol (PE, four reactive groups), polypropylene glycol (PPG, two reactive groups), and an excess of isophorone diisocyanate (two reactive groups) (Fig. 4.7). Pentaerythritol is added to the reaction at 10% by weight to yield an isocyanate-terminated polymer with a NAVG MW equal to 2720 daltons. The monomer equivalent weight of PE is 1360, obtained by dividing the monomer molecular weight by the weight percent charged (136/0.10). Pentaerythritol has four reactive alcohol sites; two form the polymer backbone and the remaining two form branches. With the use of eq. 4.7, the degree of branching for this polymer is four [(2720/1360) × (4 − 2) = 4]. The total number of end groups is (4 + 2) = 6. Each end group is assumed to be an isocyanate because of the excess isophorone diisocyanate present. The FGEW is calculated by dividing the NAVG MW by the total number of end groups theoretically present. For this example, FGEW = (2720/6) = 453.

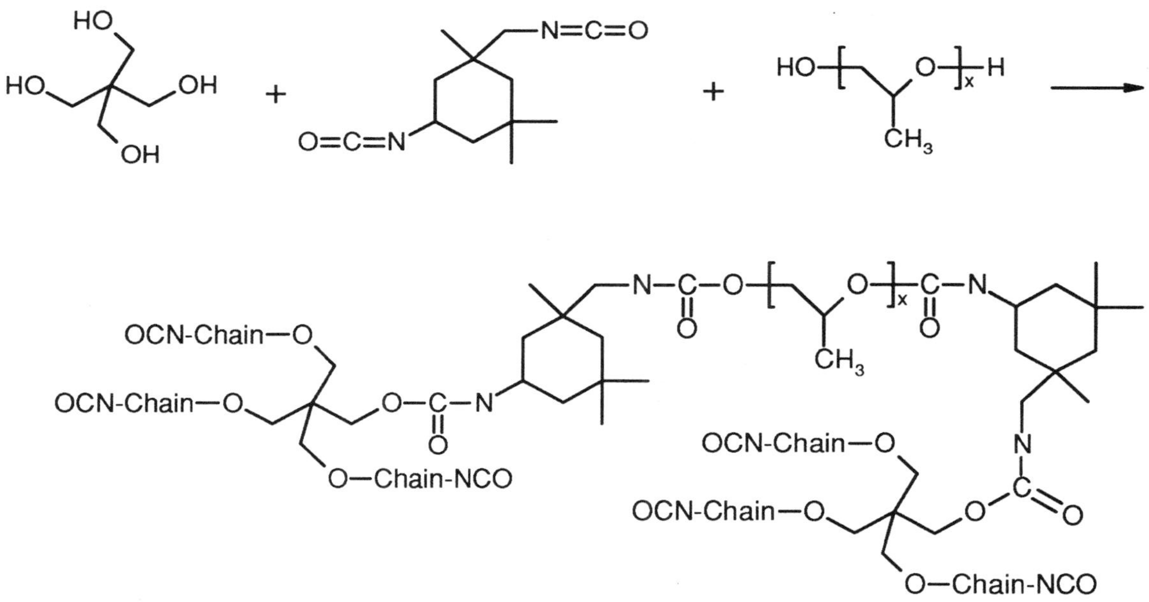

Figure 4.7 Isocyanate-terminated urethane (Example 18).

Computer programs may be useful for FGEW calculations involving condensation polymers derived from complex mixtures of feedstocks. Analytical data should be used periodically to confirm computer estimates and verify eligibility.

4.9.2   More Complex FGEW Calculations

A simple end-group analysis is insufficient to determine an accurate FGEW in cases in which some of the functional groups along the backbone of the polymer are unreacted. The equations in this section are useful for calculating FGEWs for elements, reactive groups that are unchanged under the reaction conditions, and multiple types of functional groups that remain in the polymer molecule.

Equation 4.10 is applicable to any reactive functional group in a polymer, including an atom, such as nitrogen.

$$FGEW = \frac{(FWG) \times 100}{(W\%G)} \qquad \text{(Eq. 4.10)}$$

where FWG is formula weight of the group and W%G is weight percent of the group.

Example 19

A polymer contains 2.8% by weight basic nitrogen. Using 14.0, the atomic weight of nitrogen, as the formula weight of the group, the amine FGEW may be calculated as follows:

$$FGEW = \frac{14.0 \times 100}{2.8} = 500$$

The weight percent of a functional group in a polymer may be calculated using eq. 4.11. This equation is valid as long as the monomer is included in its entirety and the functional groups are introduced unchanged.

$$\text{Weight\% of Group} = \frac{(FWG) \times (NGM) \times (W\%M)}{(FWM)} \qquad \text{(Eq. 4.11)}$$

where NGM is the number of groups in the monomer, W%M is the weight percent of the monomer, and FWM is the formula weight of the monomer.

The FGEW is obtained by substituting eq. 4.11 into eq. 4.10.

$$FGEW = \frac{(FWM) \times 100}{(W\%M) \times (NGM)} \qquad (Eq. 4.12)$$

Example 20

For an acrylic polymer containing 5.4% by weight of acryloyl chloride (formula weight = 90.5 g/mol) as a monomer, the FGEW of acid chloride groups in the polymer is 1676.

$$FGEW = \frac{(90.5)(100)}{(5.4)(1)} = 1676$$

When the moderate- and high-concern functional groups in a polymer arise from more than one monomer, the $FGEW_{combined}$ may be calculated using eq. 4.13. This equation is also useful in situations in which several different monomers contain the same functional groups. For example, if three monomers contribute epoxides that remain intact in the polymer, eq. 4.13 may be used to calculate the epoxide FGEW. The combined epoxide FGEW would be compared to the minimum permissible FGEW for epoxides in determining the eligibility of the polymer for exemption.

$$FGEW_{combined} = \frac{1}{\dfrac{1}{FGEW_1} + \dfrac{1}{FGEW_2} + \cdots + \dfrac{1}{FGEW_n}} \qquad (Eq. 4.13)$$

where $FGEW_n$ is the FGEW for each functional group in the polymer.

Example 21

In addition to end groups of interest, some condensation polymers contain unreacted reactive functional groups, such as the epoxide-capped phenol-formaldehyde novolak resin shown in Fig. 4.8. The condensation polymer is produced by the copolymerization of *para-*

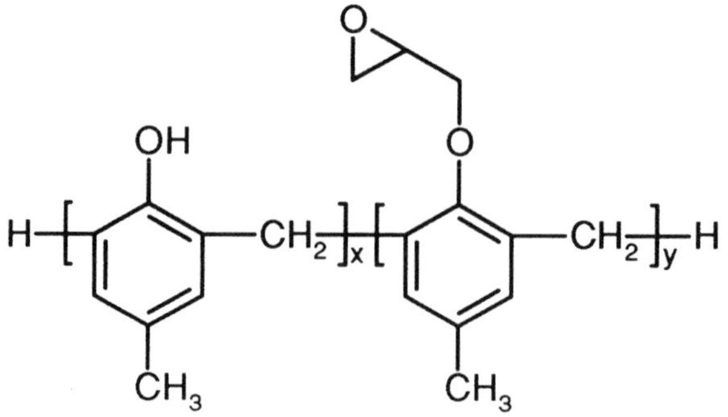

Figure 4.8   Epoxide-capped phenol-formaldehyde novolak resin (Example 21).

cresol and formaldehyde, which is reacted with 1% epichlorohydrin. The polymer is assumed to be phenol terminated as a worst-case scenario. Therefore, phenol groups with reactive *ortho* positions reside at the polymer backbone termini. GPC analysis of this polymer yields a NAVG MW of 8000 daltons. The $FGEW_{combined}$ is calculated by summing the individual FGEWs for each type of reactive group in the molecule, both end groups and unreacted groups, using eq. 4.13. The FGEW for the terminal phenolic *ortho* positions is 4000 daltons (8000/2). This value exceeds the minimum permissible functional group equivalent weight for the phenol group, which is of moderate concern (1000 daltons minimum permissible weight). If no other reactive groups were present in the polymer, then it would be eligible for exemption. However, the FGEW of epoxide must also be calculated because of the presence of the epoxy rings from the epichlorohydrin. Note that the FGEW for the epoxide must be included in the $FGEW_{combined}$ calculation, even though epichlorohydrin is not included in the chemical identity of the polymer, as it is charged at less than 2% by weight. The epoxide FGEW is calculated to be 9250 using eq. 4.12.

$$FGEW = \frac{(92.5)(100)}{(1)(1)} = 9250$$

The FGEW value means that there is one epoxide moiety for every 9250 daltons of polymer. Again, the polymer would meet the FGEW exemption criteria if epoxide, a group of moderate concern, were the only reactive functional group in the polymer. Because two types of reactive groups of concern are present, a $FGEW_{combined}$ must be calculated to determine exemption eligibility. With the use of the individual FGEWs for phenol and epoxide, the $FGEW_{combined}$ is calculated from eq. 4.13:

$$FGEW_{combined} = \frac{1}{\dfrac{1}{4000} + \dfrac{1}{9250}} = 2792$$

This polymer would be eligible for exemption because the $FGEW_{combined}$ of 2792 daltons is greater than the required 1000 minimum permissible equivalent weight (threshold level for moderate-concern groups).

Suppose that a reactant with a high-concern functional group was substituted for the epichlorohydrin in the preceding example. If acryloyl chloride (MW = 90.5 g/mol) replaces epichlorohydrin, the resulting polymer would be ineligible for the exemption. The presence of both a moderate- and high-concern functional group gives a $FGEW_{combined}$ of 2774 daltons, below the minimum permissible equivalent weight of 5000 daltons.

$$FGEW_{combined} = \frac{1}{\dfrac{1}{4000} + \dfrac{1}{9050}} = 2774$$

Example 22

Similar calculations are applicable to addition polymers. For example, the radical polymerization of acrylates proceeds through the alkene, leaving reactive functionality in the molecule. It is reasonable to assume that each monomer charged to the reaction vessel will be incorporated in its entirety to form the polymer.

Assume that polyacrylate was produced from 10% glycidyl methacrylate (MW = 142 g/mol), 2% hydroxymethyl acrylamide (MW = 101 g/mol), and 88% acrylic acid (Fig. 4.9). Two moderate-concern reactive functional groups are present: the epoxide from glycidyl methacrylate and the hydroxymethyl amide from the acrylamide. The

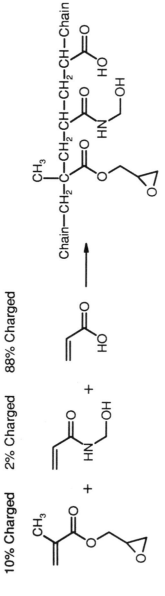

Figure 4.9 Polyacrylate with multiple functional groups (Example 22).

carboxylic acid functional group (a low-concern functional group) from acrylic acid may be used without limit.

The FGEW for the reactive functional groups may be determined using eq. 4.12. For the epoxide, the FGEW is 1420 daltons (142/0.10); for the hydroxymethyl amide, the FGEW is 5050 daltons (101/0.02). With the use of eq. 4.13, the FGEW$_{combined}$ is 1108 daltons.

$$FGEW_{combined} = \frac{1}{\dfrac{1}{1420} + \dfrac{1}{5050}} = 1108$$

This polymer would be eligible for the exemption because the 1000-dalton threshold for moderate-concern reactants was met. Because the FGEW$_{combined}$ is so close to the 1000-dalton threshold, the manufacturer will have little leeway to increase the epoxide or amide in future batches. Each batch must meet the exemption. If it is anticipated that some batches will not meet the exemption, the manufacturer or importer must file a regular PMN 90 days before the manufacture of those particular production runs.

In some addition reactions, the reactive groups engaged in the polymerization reaction are consumed (**Example 23**), whereas in others they are not (**Example 24**).

Example 23

The reaction between an amine and an isocyanate yields a polymer with an unreactive "urea" backbone. End-group analysis would be used to determine if the FGEW falls within the allowable limits for exemption.

Example 24

A polymer containing reactive functional groups in its backbone is a more complex case. Consider the reaction between ethanediamine (MW = 60 g/mol), charged at 30%, and diglycidyl ether (MW = 130 g/mol), charged at 70% (Fig. 4.10). Amine nitrogens react with the epoxides to form aliphatic alcohols, low-concern groups that may be present in any quantity. The amine functionality, however, remains

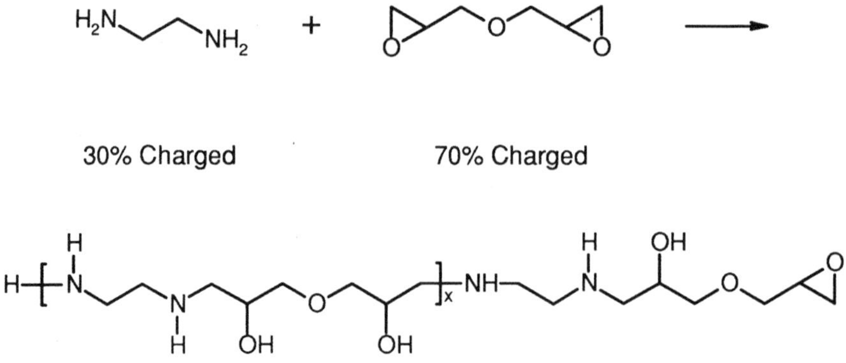

30% Charged   70% Charged

Figure 4.10   Polymerization of ethanediamine and diglycidyl ether (Example 24).

intact. The FGEW for the amine is proportional to the amount of feedstock containing the amine charged to the reaction vessel. Equation 4.12 is used to estimate the FGEW for the amines:

$$FGEW = \frac{60 \times 100}{30 \times 2} = 100$$

The minimum permissible equivalent weight for amines, a high-concern group, is 5000 daltons. The amine content renders the polymer ineligible for exemption; no further calculations are necessary because adding more groups to the $FGEW_{combined}$ calculation will only lower the value.

Some addition polymer processes employ a large excess of one reactant or group of reactants. For polymers made under these conditions, a simple repeating unit of known molecular weight can be assumed. The FGEW is obtained by dividing the unit molecular weight by the number of groups in the unit. Residual amounts of monomers or other reactants need not be reported under the new rule. The amount of reactant that does not form polymer is not regulated by the new polymer exemption rule. These residual, unreacted materials must be on the TSCA inventory and are covered as existing chemicals by different Agency authority.

Example 25

A polyamine was made by combining 70% by weight 1,2-benzene-diamine (MW = 108 g/mol) with 30% by weight diglycidyl ether (MW = 130 g/mol). A linear polymer of a 1:1 adduct (MW = 238 g/mol) is the likely repeating unit (Fig. 4.11). The amine FGEW would be 119 daltons (one-half the NAVG MW). The FGEW remains constant regardless of the number of repeating units in the polymer or the amount of excess diamine monomer charged to the reaction vessel.

4.9.3 Determining FGEW by Nomograph

The nomograph in Fig. 4.12 has been developed to aid in the estimation of FGEW. The logarithmic axes on the nomograph are "formula weight of group or monomer," "FGEW," and "weight percent of group or monomer in the polymer." The FGEW for the monomer or group can be estimated by selecting the appropriate axis points on the first and last graphs; connecting these points will intersect the FGEW graph at the approximate functional group equivalent weight. If the monomers contain several identical groups, divide the FGEW by the number of identical groups in the monomer. Equation 4.4 should be used when several different monomers contain the same groups.

4.10 OTHER REGULATIONS AND REQUIREMENTS

Please consult the new rule at 60 FR 16316–16336 (USEPA 1995; 2) for any of the following topics:

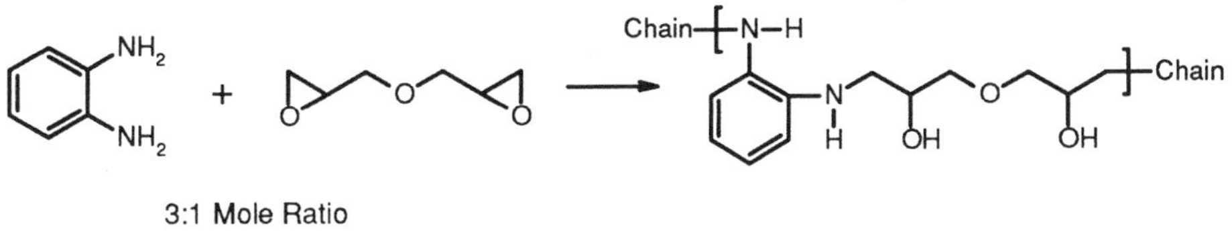

3:1 Mole Ratio

Figure 4.11 Polymerization of 1,2–benzenediamine and diglycidyl ether (Example 25).

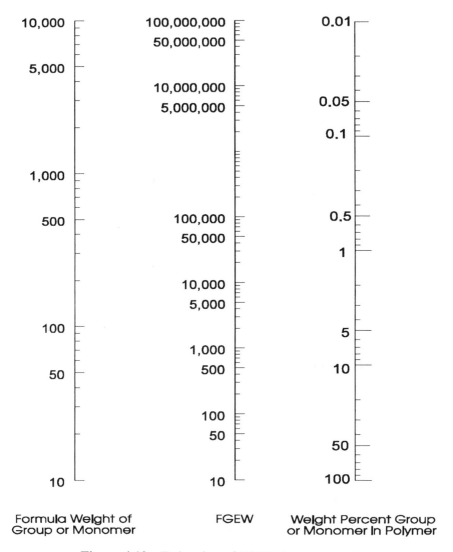

Figure 4.12   Estimation of FGEW by nomograph.

- exemption report and requirements,
- chemical identity information,
- certification,
- exemptions granted under superseded regulations,

- record keeping,
- inspections,
- submission information,
- compliance,
- inspections, and
- confidentiality.

4.11   FREQUENTLY ASKED QUESTIONS

Polymer Definition

1. Should I use the "new" polymer definition to determine whether a polymer is on the Inventory? For example, if I manufacture an alkyl ethoxylate sulfate, $R(OCH_2CH_2)_nOSO_3Na$, where n = an average of seven, will I have to submit a PMN even though more than three consecutive monomer units are present? What if n = exactly 7 or exactly 15? The Inventory currently considers all the ethoxylates with greater than three units as polymeric and therefore as the same substance.

The alkyl ethoxylate sulfates with definite numbers of repeating units would not meet the polymer definition because they would consist of molecules of a single molecular weight. However, Chemical Abstracts' nomenclature rules and the TSCA Inventory treat *some* of these as though they were polymers. For example, "laureth sulfate," where $R = C_{12}H_{25}$ and $n = x$, is on the Inventory (CASRN 9004-82-4). Variations in the number of ethylene oxide units will not produce a new (non-Inventory) substance as long as n is either >10 or variable or represents an average value. Laureth sulfate, with n averaging 7, is considered an existing substance as is laureth sulfate, with n = exactly 15. If n = exactly 7, however, the substance is not considered a polymer but a discrete chemical entity. It would have a different name and Chemical Abstracts Service Registry Number (CASRN) and would be a new chemical, if not already listed elsewhere on the Inventory.

The "new" polymer definition does not affect the Inventory status of existing polymers or of new polymers submitted under the PMN rule. The polymer definition applies only to polymers manufactured under the polymer exemption.

2. Would the following example count as a "polymer molecule"? The longest straight chain is 1 + 1 + 2 = 3 + 1.

H(oxypropylene)–O–sorbitol–O–(propyleneoxy)$_2$–H

No. Sorbitol cannot be a repeating unit under the conditions of the polymerization reaction (propoxylation) and is considered an "other reactant." Therefore, the longest sequence of monomer units (derived from propylene oxide) is two. A "polymer molecule" requires a continuous string of at least three monomer units, plus one additional monomer unit or other reactant.

3. How do you apply the molecular weight distribution requirement of the polymer definition (i.e., <50% of any one MW) to highly cross-linked polymers of essentially infinite MW?

Unless the entire mass of polymer produced was in one continuous phase, the actual molecular weight would be limited by the size of the individual droplets, beads, pellets, flakes, and so forth. No two of these would likely have exactly the same mass, and the distribution criterion would be met.

Elemental Exclusions

4. Regarding elemental limitations, why was fluorine not included in 723.250(d)(2)(ii)(B) but included in (ii)(C)?

Fluoride ion (F$^-$) has a high acute toxicity and is unacceptable as a counterion in a substance that is supposed to present no unreasonable risk to human health. Fluorine covalently bound to carbon is either unreactive and not available in the ionic form (F$^-$) or is part of a reactive functional group such as acyl fluoride (COF) and subject to the reactive functional group criteria.

5. Can you give an example of F$^-$ (anion) that is not allowed?

A cationic ion exchange resin that is eligible for exemption based on insolubility would be ineligible if the counterion is fluoride (F$^-$).

6. Ammonium is not listed as an acceptable monatomic counterion. Does this mean a polymer may be made under the exemption but not its ammonium salt?

No. Ammonium may be used as a counterion; it is not monatomic and is not excluded under section (d)(2)(ii).

7. Are only monatomic counterions allowed? What about polyatomic counterions, such as CO_3^{2-}, HCO_3^-, and NO_3^-?

Only those monatomic counterions found under section (d)(2)(ii)(B) are allowed. All other monatomic counterions are excluded. The polymer exemption rule places no specific restrictions on polyatomic counterions. They are permitted as long as they do not otherwise render the polymer ineligible. For example, carbonate (CO_3^-) is allowed but perchlorate (ClO_4^-) is not because the chlorine is neither monatomic nor covalently bound to carbon. Trichloroacetate ($CCl_3CO_2^-$) is allowed because the chlorine is covalently bound to carbon.

8. Are monomers that have CF_2 or CF_3 groups allowed?

Monomers that contain CF_2 or CF_3 groups are acceptable, provided they are not part of a reactive functional group. —CF_2— is not generally a monomer unit because it is not "the reacted form of the monomer in the polymer." However, —CF_2CF_2— groups derived from the polymerization of tetrafluoroethylene certainly could be monomer units.

Exclusion for Degradable Polymers

9. What is the time frame for "polymers that do not degrade, decompose, or depolymerize"? Does EPA want us to synthesize polymers that bioaccumulate in the environment? Does the term "degrade" apply to biodegradation or other degradation in waste treatment systems?

This restriction is essentially unchanged from the 1984 polymer exemption. Although the EPA recognizes in principle the beneficial effects of biodegradability, it commented in the discussion section of that rule that the Agency "has little experience reviewing the mechanism by which breakdown may occur, the decomposition products that may result, and the potential uses of such polymers. . . . Because of the complexity of review necessary for many of these polymers and the lack of EPA review experience, the Agency did not believe that an expedited review period was sufficient to adequately characterize risk."

In that discussion, the Agency acknowledged that essentially all polymers degrade or decompose to a limited degree over time. It gave as examples the normal fate of polymers in landfills and the weathering of paint and specifically stated that the exclusion was not intended to address such degradation. *Substantial* biodegradation in a waste treatment system would render a polymer ineligible for the exemption.

10. How does the EPA define "degrade," "decompose," and "depolymerize"? Assuming other criteria are met, can a polymer still be eligible for the exemption if degradation, decomposition, or depolymerization occur as a by-product minor reaction?

The definition may be found at §723.250(d)(3): "For the purposes of this section, degradation, decomposition, or depolymerization mean those types of chemical change that convert a polymeric substance into simpler, smaller substances, through processes including but not limited to oxidation, hydrolysis, attack by solvents, heat, light, or microbial action." Minor by-product degradation will not exclude a polymer from the exemption.

11. Starch is a polymer that readily degrades in the environment. Would starch be eligible for the exemption if it were not listed on the TSCA Inventory?

No; polymers that readily degrade are excluded from the exemption.

12. What does the Agency mean by "substantially" in the phrase "substantially degrade"? Does this term refer to specific conditions or to normal environmental conditions?

"Substantially" refers to degradation that occurs to a significantly large extent. The restriction refers to polymers that undergo considerable degradation under normal conditions of use or disposal, in a reasonable length of time.

13. If a polymer is designed to be pyrolyzed or burned when it functions as intended, is that polymer excluded from exemption by the "degrade, decompose, or depolymerize" conditions?

Yes, if that is the normal way it is used. For example, a polymer propellant or explosive would be excluded. A polymeric flame retardant that functions by decomposing when exposed to fire would be excluded. A plastic used for garbage bags would not be excluded simply because it might be incinerated.

14. A manufacturer produces a polymer that is readily biodegradable by the OECD test; otherwise, it is eligible for the polymer exemption. There are two uses for the product. In the first case, the manufacturer can reasonably anticipate that the polymer will ultimately reside in an aqueous environment, where it may degrade. In the second case, the polymer will be formulated at a low percentage into articles such that the articles themselves would not be anticipated to degrade once they are disposed of in a landfill. Provided the manufacturer could control customer sales to ensure that the polymer would only be used in the second case, could the polymer exemption apply?

Yes, provided that the use is restricted to conditions under which the polymer would not be expected to degrade, decompose, or depolymerize.

15. Will the EPA specify testing conditions for evaluating "degradation"? Will manufacturers using the exemption be required to prove that their polymers don't degrade? Can we rely on *intent* to degrade?

The Agency does not specify test conditions for degradability. There is no testing requirement to establish nondegradability. As stated in section (d)(3), polymers are excluded "that could substantially decompose after manufacture and use, even though they are not actually intended to do so." In other words, the intent of the manufacturer does not determine whether this criterion is met. A polymer is excluded if it can be reasonably anticipated that it will substantially decompose.

16. Are Diels–Alder polymers (for example, dicyclopentadiene polymers) considered degradable?

If Diels–Alder polymers are "designed or reasonably anticipated to substantially degrade, decompose, or depolymerize," they would be excluded; if not, they would be eligible if the other exemption criteria are met. There are no specific constraints on structure or method of polymerization.

Exclusion of Water-Absorbing Polymers

17. How are water-soluble, water-dispersible, and water-absorbing polymers distinguished with regard to the polymer exemption? Are they treated the same? Is dispersibility considered degradation?

Water-soluble and water-dispersible (self-dispersing or already dispersed) polymers are not considered to be water-absorbing substances. Only water-absorbing, water-insoluble, nondispersible polymers are excluded. The distinction is based on an assumed mechanism for lung damage by water-absorbing polymers, which involves a failure of the lungs to clear particles of these materials. Water-soluble or water-dispersible materials are expected to be cleared and are thus not excluded. Dispersibility is not considered to be degradation.

A water-absorbing polymer that is manufactured or imported in water and is sold in water at concentrations allowing full water absorption is not excluded from exemption provided that it meets all other exemption criteria and is not otherwise excluded.

18. Why are high molecular weight water-absorbing polymers excluded from the polymer exemption?

The EPA excluded this category of polymers from the exemption based on TSCA section 8(e) inhalation study, designated 8(e)-1795 and FYI-470. A potential cancer concern was indicated for this type of high molecular weight water-absorbing polymer, specifically a water-absorbing polyacrylate polymer with a MW in excess of 1 million daltons. The Agency concluded that exposure to respirable fractions of these polymers might present an unreasonable risk to human health. (See pages 16319–16320 in the rule for a discussion of this issue.)

19. A polymer is partly ionized by a pH change, increasing its water absorption to greater than 100% by weight. Is the polymer still eligible for the exemption? What if the neutralizing agent is less than 2% by weight of the polymer? Does the so-called "(h)(7)" pH neutralizer exclusion apply to polymers >10,000 MW that absorb more than 100% of their weight of water on neutralization? Which takes precedence, the "(h)(7)" exclusion or the polymer exemption?

If the polymer becomes water absorbing on use in neutral water, it is a water-absorbing polymer, whether or not ionization is involved.

If the polymer is deliberately converted to a water-absorbing polymer by neutralization, that constitutes manufacture for commercial purposes as a chemical substance per se, rather than processing. The resulting substance would be a *different* polymer that would be considered water absorbing and would not be eligible for exemption. Even if the neutralizing agent is less than or equal to 2%, a polymer must still meet the eligibility requirements to be exempt. The unneutralized starting polymer could still be eligible for the exemption if it met the other exemption criteria.

If the neutralization results in a substance excluded from reporting under 40 CFR §720.30(h)(7) (which basically covers processing rather than manufacture), that substance remains excluded from reporting even if it would have been ineligible under the polymer exemption. (The Agency's clarification of this issue may be obtained through the TSCA Assistance Information Service, 202-554-1404, by requesting the package from the Deputy Director, Office of Pollution Prevention and Toxics to the regulated community, June 29, 1994.) If an exempt polymer is converted to a water-absorbing substance by a chemical process or reaction that produces a substance excluded from reporting under (h)(7), the starting polymer remains exempt. Both the polymer exemption and 40 CFR §720.30(h)(7) apply independently to the respective substances.

20. I have an acidic resin that is eligible for a polymer exemption. Would the salt of this resin automatically be eligible for exemption?

A salt of an exempt polymer would not automatically be eligible for exemption. If the conversion of the resin to its salt introduces no properties or constituents that would cause it to be excluded from the exemption, then the resulting polymer salt should also be eligible for the exemption. The salt, for example, would be excluded if it were a water-absorbing polymer or if it contained certain elements in amounts greater than permitted. The manufacturer must ensure that the polymer salt meets all requirements for exemption and that conversion to the salt has not caused changes in the polymer that could exclude it from exemption.

Limitation on Cationic Properties

21. Would a very "nonbasic" amine, such as a dialkyl aniline, be expected to become cationic in the environment? Suppose that the pK_a of the amine and the concentration of amine groups in the polymer allow you to calculate that the functional group equivalent weight of the protonated form of the amine will be >5000 in a natural aquatic environment. Could the polymer be eligible for the exemption?

If a manufacturer or importer can establish by pK_a data or via another method that the amine groups in a polymer are "nonbasic" and would not become cationic in the environment, the polymer would not be excluded from exemption on the basis of potentially cationic character. However, amine groups are still considered reactive functional groups whether they are protonated or not. The functional group equivalent weight is not affected by the pK_a or the "nonbasic" character of the amine.

22. Does the phrase "used only in the solid phase" refer to end use as opposed to processing, in which the polymer may be melt extruded, injection molded, and so forth?

"Used only in the solid phase" does refer to end use. A solid material melted during processing is not considered a liquid if it is solidified at the end of the processing step.

23. A polymer contains a potentially cationic group. The polymer is neither water soluble nor water dispersible but is manufactured by emulsion polymerization and therefore exists as particles dispersed in water. Is the polymer ineligible for the exemption?

Cationic and potentially cationic polymers (see Section 4.4.1) are excluded from the exemption, except for two types: 1) those that are solids, neither water soluble nor water dispersible, used only in the solid phase, and not excluded by other factors; and 2) those that have low cationic density and are not excluded by other factors. If the polymer is neither water soluble nor water dispersible, manufacture by emulsion polymerization alone would not render it ineligible.

24. What is meant by water insoluble with respect to cationic polymers that qualify for exemption? Is a specific test recommended by the phrase "[T]he polymer is a solid material that is not soluble or dispersible in water"?

The phrase in section (d)(1)(i) does not relate to a specific test. The Agency has not prescribed a specific test for water solubility of polymers. The solubility criterion, however, should be applied to the commercial material as manufactured and sold. For example, an aqueous emulsion is a water-dispersed material and a substance in that form would be considered soluble or dispersible; therefore, it would not qualify for the exemption.

Reactive Functional Groups

25. Are amine salts permitted? Are sulfonic and sulfuric acids (—SO₃H and —OSO₃H) and their salts considered nonreactive?

Amine *counterions* are permitted for anionic polymers. Sulfonate salts are not considered reactive, but sulfonic and sulfuric acids are considered reactive. The 1984 polymer exemption rule designates sulfonic and sulfuric acids as reactive, and the interpretation remains the same in the new rule.

26. Please provide a few examples of "high-concern" and "low-concern" functional groups as specified in the (e)(1) criteria. Would acrylate, epoxide, or isocyanate groups be classified as "high-concern" or "low-concern" groups?

Epoxides are found in (e)(1)(ii)(B), the list of "moderate-concern" groups for which concern exists at a functional group equivalent weight of 1000 or less. Acrylate and isocyanate are not listed as "low-concern" [(e)(1)(ii)(A)] or "moderate-concern" [(e)(1)(ii)(B)] functional groups; therefore, they are considered "high-concern" groups and fall under (e)(1)(ii)(C). High-concern functional groups have a minimum functional group equivalent weight threshold of 5000 daltons. Any reactive group not listed in sections (e)(1)(ii)(A) or (e)(1)(ii)(B) is considered to be "high concern."

27. The nitro group does not appear on the low- or moderate-concern list of reactive functional groups. Does the nitro group fall into the high-concern category by default? This is counterintuitive because I wouldn't consider the nitro group to be very reactive and of much concern.

Numerous groups, such as ester and ether, are not listed because they are not considered to be reactive functional groups. Nitro groups are also not considered reactive functional groups unless they are specially activated. For example, certain aromatic nitro groups are readily displaced in nucleophilic substitution reactions.

28. Is the amine group considered a high-concern reactive functional group? It is not listed at either 40 CFR §723.250(e)(ii)(A) or (B), which by default places it in category (C). However, the criterion for a substance that "may become cationic in the environment" appears to address the concerns that EPA has for amine groups by limiting the amount of amine in a polymer to one in 5000 daltons. It seems that the amine group, in and of itself, should not be regarded as a reactive functional group. Would the amine group be used in the calculation for FGEW$_{combined}$?

The amine group is considered a high-concern reactive functional group and is used in calculating FGEW$_{combined}$. It is, for example, reactive in condensation reactions that yield polyamides and polyimides. The Agency's concern for amines as a reactive functional group extends beyond its considerations of aquatic toxicity. For polymers that are not water soluble or water dispersible and used only in the solid phase, the limitation on cationic functional groups does not apply, but the limit on amine groups as reactive groups still does apply.

29. How does one determine the FGEW for a polymer containing both "high-concern" and "moderate-concern" groups? Are the high-concern groups combined separately from the moderate-concern groups, or are both added together? Do low-concern groups enter into the calculation?

If "high-concern" [(e)(1)(ii)(C)] groups are present, the combined functional group equivalent weight is calculated from both the "high-concern" and "moderate-concern" [(e)(1)(ii)(B)] groups present in the polymer. The FGEW$_{combined}$ must be greater than or equal to 5000 daltons to meet the eligibility criterion. "Low-concern" [(e)(1)(ii)(A)] groups are not included in the FGEW$_{combined}$ calculation.

30. If a polymer with a number-average molecular weight >10,000 daltons meets the reactive functional group and oligomer content criteria of (e)(1), but not the more stringent oligomer content criterion of (e)(2), doesn't it fall into a gap between (e)(1) and (e)(2)? Is this polymer eligible for exemption? If not, does the Agency plan to amend the (e)(1) criterion to omit the phrase "and less than 10,000 daltons"?

The (e)(1) and (e)(2) exemptions are mutually exclusive. Polymers with molecular weight of more than 10,000 are eligible only for the (e)(2) exemption, which has lower allowable concentrations of oligomer than (e)(1). The polymer described above would not be eligible for either the (e)(1) or (e)(2) exemption. The Agency received no comment on this issue from the time it was proposed on February 8, 1993, until after the final rule became effective on May 30, 1995. A modification of the criteria seems reasonable, but additional rulemaking would be required. The issue is under discussion.

The "Two Percent Rule" (and Non-Inventory Reactants)

31. Please explain the changes in the "Two Percent Rule" for polymers.

The "Two Percent Rule," in effect since 1977, allows manufacturers and importers of polymers to add monomers or other reactants to an Inventory-listed polymer at levels of 2% or less (based on the dry weight of the manufactured polymer), without making a polymer with a different chemical identity than the Inventory-listed polymer. The "Two Percent Rule" also serves as a basis for determining the identity of a polymer. Before the PMN rule amendments became effective on May 30, 1995, the monomer content of a polymer was calculated on the basis of weight percentage of monomer or other reactant "charged" to the reaction vessel. The 1995 amendments allow greater flexibility in determining the percent composition and whether monomers and other reactants are present at more than 2%. In addition to the "charged" method, the 1995 amendments permit determination of the amount of monomer or other reactant present using the "incorporated" method. Using this method, manufacturers and importers report the theoretical minimum weight percent of monomers or other reactants needed to account for the amount present "in chemically combined form" (incorporated) in the polymer. Either method may be used to determine the 2% level. Although the incorporated method provides more flexibility, analytical data or theoretical calculations are required to support this technique.

This change in the "Two Percent Rule" applies to all polymers under TSCA, including Inventory listings, PMN submissions, and polymer exemptions.

32. Using the "incorporated" method, what records are required to support the claim that 2% or less of a reactant are incorporated in a polymer, even though a higher level of reactant is charged to the reaction vessel?

Ideally, records will provide analytical data to demonstrate that the minimum weight of monomer and/or reactant required to account for the monomer and/or reactant fragments chemically incorporated is 2% or less. If such an analysis is not feasible, appropriate theoretical calculations must be provided. Potential variations from batch to batch should be considered.

33. It appears from the polymer exemption rule and the technical guidance manual that a reactant and/or monomer present in a polymer at less than or equal to 2% may not be included in the identity of the polymer. Is this true?

Yes; if a polymer has less than or equal to 2% of a monomer and/or reactant, the identity does not contain that monomer/reactant. If an otherwise identical polymer is made containing the same monomer and/or reactant at greater than 2%, the identity of the second polymer is different from the first. Two exemptions would have to be claimed to cover both polymers because the "identity" is established by the percentages of monomers and/or reactants charged or incorporated in the polymer.

For polymers for which a PMN is submitted, the submitter does have the option of including a reactant and/or monomer at less than or equal to 2% in the polymer identity.

34. Does a manufacturer need to test every batch of polymer to prove that less than 2% of a monomer and/or reactant is incorporated, or would one documented test on a typical batch be sufficient?

A company is not required to test every batch but is required to maintain in its records analytical data or theoretical calculations to demonstrate compliance with the "Two Percent Rule" when using the "incorporated" method. If occasional batches are expected to exceed that level, the manufacturer should employ more frequent testing, always consider the reactant to be part of the chemical identity, or manufacture a separate exempt polymer with the reactant incorporated at greater than 2% and included in the polymer identity.

35. I use a prepolymer that is on the Inventory to make my polymer. The prepolymer contains a non-Inventory monomer, and the final polymer contains greater than 2% of that monomer. Is my polymer ineligible for the exemption?

The polymer is not ineligible for the exemption based on the presence of the non-Inventory monomer. Although §(d)(4) bars the use of "monomers and/or other reactants(that are not already on the TSCA Chemical Substance Inventory," it is the prepolymer that is the reactant and is listed on the Inventory. However, the identity of the final polymer will probably include the non-Inventory monomer (see questions 45–49 for related topics).

36. If an initiator is incorporated at no more than 2%, does it have to be on the TSCA Inventory?

An initiator or other reactant present at no more than 2% does not have to be on the Inventory for a polymer to be eligible for the exemption. If the reactant is not on the Inventory, however, it cannot be *used* for commercial manufacture in the United States. Consequently, this provision is applicable only to imported polymers.

37. Can I use any monomer on the Inventory at less than or equal to 2%?

Yes, as long as that monomer doesn't introduce elements, groups, or properties that would render the polymer ineligible at the concentration of monomer used. Note that the (e)(3) "polyester" exemption requires *all* components of the polymer to be on the list of allowable reactants. A polymer would be ineligible for the (e)(3) exemption if nonlisted monomers were used, even at 2% or less.

38. I wish to import a polymer containing greater than 2% of a reactant not on the public TSCA Inventory but which may be on the confidential inventory. Can I file a Notice of Bona Fide Intent to Manufacture to determine if the polymer meets the polymer exemption criteria or do I need to file a PMN for the polymer?

There is really no way to find out whether a substance is on the Inventory unless you intend to import or manufacture that substance itself. You may not file a "Bona Fide" on the reactant unless you have a bona fide intent to manufacture or import it. (Your supplier, if in the United States, could file a Bona Fide on the monomer.) Therefore, the only substance for which you can file a Bona Fide is the final polymer. No PMN is needed if the polymer is on the Inventory. If not, you will need to file a PMN for the polymer. Unless you have a real intent to import or manufacture the monomer, you cannot file a PMN or an exemption for the monomer. Your polymer may be eligible for exemption if the monomer is on the Inventory. If it is not, completing the review process for the monomer and commencing its manufacture or import will allow it to be used in an otherwise exemptible polymer.

39. Does the polymer exemption apply to an imported polymer made with a non-TSCA listed chemical? If no, why not?

If the reactant is present at less than or equal to 2% and if its presence does not otherwise render the polymer ineligible, the polymer may be imported. The polymer cannot be imported under the polymer exemption if the non-TSCA reactant is used at greater than 2%. The Agency cannot determine if an unreasonable risk is posed by a polymer containing residual amounts of a monomer or other reactant that it has never reviewed. A polymer may not be manufactured domestically if any reactant is not on the Inventory.

40. If a polymer is on the Inventory but contains a non-Inventory monomer, can you import it?

Yes. A polymer that is on the Inventory is an existing chemical, and no PMN or other notice or exemption is required. The exclusion of non-Inventory monomers and other reactants applies only to the polymer exemption. A polymer may not be manufactured domestically unless all the reactants are on the Inventory.

41. What if the non-Inventory-listed monomer is charged or incorporated at less than or equal to 2%?

A polymer containing a non-Inventory-listed monomer at less than or equal to 2% may be eligible for the exemption provided that the monomer does not "introduce into the polymer elements, properties, or functional groups that would render the polymer ineligible for the exemption." Although language at §(g)(1) states that such reactants are not allowed "at any level," reactants used below certain levels are allowed, provided the reactants do not render the polymer ineligible for exemption *at that use level*. Note again that a non-Inventory-listed monomer that is not on the list of permitted reactants for the (e)(3) exemption will render it ineligible for that exemption. There are, in fact, reactants on that list that are not on the Inventory. These are not subject to the 2% limitation because they have already been reviewed by the Agency and are considered to be not of concern; see the answer to question 50. However, if a monomer or other reactant is not on the Inventory or otherwise excluded from reporting or exempted from section 5 requirements, it cannot be used for domestic manufacture, regardless of its concentration in the product polymer.

42. Can polymers that utilize less than or equal to 2% of non–Inventory-listed monomers be eligible for the exemption?

Such polymers would be eligible for exemption as long as they meet all the other exemption criteria. However, a monomer used at any concentration must be on the Inventory or exempt before it can be used in the domestic manufacture of the polymer.

43. A polymer containing any amount of a component that is not on the TSCA Inventory cannot be manufactured domestically under the polymer exemption. Does this mean that a PMN for the polymer is necessary, or does it mean that the reactant must first be put on the Inventory before the polymer exemption can be used?

To use a substance domestically for any reason, it must be on the Inventory, excluded from reporting, or exempted under an applicable section 5 exemption (for example, low volume, low release and exposure, pre-1995 polymer, or current polymer). Therefore, a PMN (or applicable section 5 exemption) is required for the new reactant, and the reactant must be on the Inventory or exempt before it can be used in the domestic manufacture of the polymer. Once the reactant is on the Inventory, a polymer containing it is not automatically excluded from the exemption, as long as it is otherwise eligible.

44. If a TSCA-listed brominated flame retardant is mixed at greater than 2% in a polymer base, is the polymer subject to PMN requirements or is it exempt?

The material is considered to be a mixture of polymer and the flame retardant. Mixtures are not subject to reporting under TSCA provided there is no intended reaction between the components of the mixture. The components of the mixture are subject to reporting if they are not on the Inventory. If the polymer is eligible for exemption, the presence of the other component will not render it ineligible.

45. Are all of the exclusions under 40 CFR §720.30 ("Chemicals not subject to notification requirements") applicable to the polymer exemption?

Yes; however, a manufacturer must comply with the conditions of the exclusions even though the substances are being used in connection with the polymer exemption. For example, a substance subject to the low-volume exemption could be used as a monomer for an eligible polymer, but only if the supplier is a holder of the exemption and if the appropriate production ceiling is observed.

Inventory Status of Reactants; Chemical Identity of Polymers

46. How do I find out whether

(a)   my polymer is on the confidential TSCA Inventory?

(b)   a reactant in my polymer is on the confidential Inventory?

You can determine the Inventory status of your polymer by filing a Notice of Bona Fide Intent to Manufacture (or a PMN). You may not file a Bona Fide on the reactant unless you have a *bona fide* intent to manufacture or import it. It is the responsibility of the manufacturer or importer of the reactant to determine its Inventory status.

47. A prepolymer is used as a precursor in the manufacture of a polymer. Are the constituents of the final polymer the ultimate reactants from which the prepolymer was made or the prepolymer itself?

The choice should follow Chemical Abstracts (CA) nomenclature rules and conventions for its Ninth Collective Index (9CI). In general, polymers are named on the basis of their ultimate monomers. The name of a prepolymer derived from dimethyl terephthalate and 1,4-butanediol would be based on those reactants. However, there are exceptions to this generalization. For example, although polyethylene glycol may be considered a homopolymer of ethylene oxide, it is not named as a homopolymer under CA nomenclature practices. Polyethylene glycol is named according to the structural repeating unit (SRU) and end groups present: α-hydro-ω-hydroxy-poly(oxy-1,2-ethanediyl). Polydimethylsiloxane is also named on the basis of its SRU (di-Me siloxanes and silicones) and is considered to be end capped with trimethylsilyl groups. If a prepolymer is named to represent a certain structural feature or definite repeating unit, its name cannot be decomposed into ultimate monomers in order to name the final polymer. The Agency's conventions for representation of polymeric substances are discussed in greater detail in the 1995 paper *Toxic Substances Control Act Inventory Representation for Polymeric Substances* available from the TSCA Hotline [202-554-1404 (phone); 202-554-5603 (fax)].

48. Does the "Two Percent Rule" apply to the actual reactants used or to the ultimate or putative reactants?

The ultimate reactants should be the basis of the chemical identity of the polymer. For example, if a new polymer is made from the prepolymer in the preceding answer (dimethyl terephthalate and 1,4-butanediol) plus additional dimethyl terephthalate and ethylene glycol, the final polymer name would be based on the three constituents (the total amount of dimethyl terephthalate would be the sum of the separate contributions). Ultimate reactants that contribute no more than 2% by weight to the final polymer may be omitted from the identity. If a homopolymer is used as a prepolymer constituent, the identity of the final polymer should be based on the ultimate monomer, except when CA practice differs due to the applicability of SRU nomenclature. Although *calculation* of the percent composition of a polymer may be based on analysis, the *identity* should be based on the ultimate precursors.

49. The modified "Two Percent Rule" allows reporting of polymers as incorporated as well as charged. Can all polymer listings on the Inventory now be read either as incorporated *or* as charged?

Yes, polymers on the Inventory may be interpreted either as incorporated or as charged. Remember that "incorporated" means the minimum theoretical amount needed to be charged to account for the amount of monomer or reactant molecules or fragments found in the polymer itself.

50. If I import a polymer that is described as a sodium salt and I can determine analytically that sodium is present at 2% or less, can I assume that sodium hydroxide was used as the neutralizing agent? Should I use the molecular weight of sodium hydroxide in

determining the percent incorporated (and hence the chemical identity)?

Yes; in the absence of information about the source of the sodium ion, sodium hydroxide should be used as the default source, and the calculations should be based on the molecular weight of sodium hydroxide. The hydroxides of magnesium, aluminum, potassium, and calcium should also be used as the default sources of the respective ions.

Polyester Criterion

51. Some of the reactants on the polyester list are not on the TSCA Inventory. Am I allowed to use these to manufacture a polyester under the polymer exemption?

Yes, for imported polymers. Under the 1984 exemption, those reactants were placed on the polyester ingredients list, even though they were not on the Inventory, because there was no exclusion for non-Inventory reactants. The Agency continues to allow these specific reactants because the Agency has determined that no unreasonable risk will be incurred by a polymer that contains residual amounts of these reactants. For domestic manufacture, you may use only substances that are on the Inventory or are otherwise exempt or excluded from reporting.

52. If a monomer in my polyester is used at less than or equal to 2% and is not on the (e)(3) list, is the polymer eligible for the exemption if it meets all other criteria and is not otherwise excluded from the (e)(3) exemption?

No, the polymer would not be eligible for the exemption. Only monomers and reactants on the (e)(3) list may be used for this category of polymer regardless of the percentage charged or incorporated.

53. Is there a mechanism to add new reactants to the polyester reactants list? If so, what is required?

The list of permissible ingredients in the present exemption has already been enlarged since the 1984 version. The Agency's response to a comment addressing this specific issue is found in the preamble to the final rule: "The Agency believes that it would be appropriate in the future to propose amendments to this section to allow expansion of the list of eligible precursors, when additional candidates have been identified. To support requests for additional reactants, petitioners should provide health and environmental effects information on the candidate reactants, which must be already on the Inventory." No specific mechanism is yet in place. The Agency would prefer not to deal with such reactants piecemeal but rather as part of a systematic process, perhaps initiated by trade organizations or consortia of interested companies.

Miscellaneous Issues

54. If a polymer contains a gel fraction (MW >10,000) of 10–20% and the MW of the soluble fraction is <10,000, is it no longer exempt? Is the gel fraction an impurity or a by-product?

Because the two polymeric fractions have the same chemical identity and are not separately prepared, they are normally considered as a single substance for which one number-average molecular weight is measured. However, impurities are not considered part of the polymer composition; if the 10–20% gel portion is undesirable, it may be considered an impurity. In this case, the appropriate number-average molecular weight would be for the portion below 10,000, and the polymer would have to meet the (e)(1) criteria. Whether the gel portion is considered an impurity does not depend on whether it is a minor component but on whether it is intended to be present.

55. Are Inventory-listed monomers that have allowed groups and a 5 (e) order attached eligible for the new polymer exemption?

Yes, as long as the use of the monomer is in accordance with the conditions of the 5 (e) order.

56. There is no guidance on measurement of oligomer content. Is accumulated weight fraction on a GPC trace an adequate determination? In the absence of GPC, how can this be done?

The Agency has not prescribed any analytical methodology. Cumulative weight fraction is a commonly accepted method; other methods may be acceptable, depending on circumstances.

57. Do polymers made by "reactive processing" of two or more other polymers (listed on TSCA) fall under the polymer exemption?

Yes, as long as they meet the necessary criteria and are not otherwise excluded. There is no exclusion for polymers made from other polymers, nor is there any restriction on method of preparation.

58. What are the analytical requirements with respect to insoluble polymers? Is inference from melt flow data and comparison to other polymers adequate? Can I use Monte Carlo simulation methods (such as Oligo 5) to provide a theoretical MW for an insoluble polymer?

The Agency does not require any specific analytical methodology. Inference from physical behavior, from comparison to close analogues, and from theoretical calculation is acceptable where appropriate or where other methods are inapplicable. Although widely used, Monte Carlo methods have not been subjected to much experimental verification. If your polymer is expected to have values of MW or oligomer content near the allowable thresholds, you should not rely too strongly on such methods.

59. Does EPA prescribe a specific analytical method for determining the amount incorporated in a manufactured polymer when using the "chemically combined" method?

No, the rule does not specify a particular method.

60. If you make a new polymer in the laboratory that meets the exemption rule, do you need to send a research and development letter to the customer?

Substances considered to be research and development (R&D) chemicals are subject to the Research and Development Exemption and must follow the conditions of that exemption. Polymers should be handled according to the R&D requirements until they reach the stage of being commercial products eligible for the polymer exemption. When the commercial activity is no longer R&D, provisions of that exemption no longer apply.

REFERENCES

1. Code of Federal Regulations. 40 CFR Chapter I, Subchapter R, part 723.250.

2. U.S. Environmental Protection Agency. 1995 (March 29). Premanufacture Notification Exemptions; Revisions of Exemptions for Polymers; Final Rule. 40 CFR Part 723 (60 FR 16316–16336).

3. Organization for Economic Co-operation and Development (OECD). 1994 (May). *OECD Guidelines for the Testing of Chemicals, Determination of the Low Molecular Weight Polymer Content* (draft proposal).

4. Organization for Economic Co-operation and Development. Chemicals Group and Management Committee. 1994 (May 10). *Chairman's Report, Third Meeting of OECD Experts on Polymers, Tokyo, 14–16 April 1993.*

5. *Encyclopedia of Chemical Technology* (4th ed.). 1996. New York: John Wiley & Sons, vol. 19.

6. IUPAC Physical Chemistry Division, Engl. 1976. *Pure Appl. Chem.* 48(2): 241–246.

7. Glover, C. A. 1975. *Tech. Methods Polym. Eval.* 4(pt. 1): 79–159.

8. Tung, L. H., and J. R. Runyon. 1973. *J. Appl. Polym. Sci.* 17(5): 1589–1596.

9. Wagner, H. L. and P. H. Verdier. 1978. *J. Res. Natl. Bur. Stand. (U.S.)* 83(2): 179–184.

10. Glover, C. A. 1973. *Adv. Chem. Ser.* Volume date 1971, no. 125, 1–8.

INDEX